BRITISH GEOLOGICAL SURVEY

J D PEACOCK
J R MENDUM and
D J FETTES

CONTRIBUTOR
D Gould

Geology of the Glen Affric district

Memoir for 1:50 000 geological sheet 72E (Scotland)

COUNTY LIBRARY
19/20 WESTMINSTER HOUSE
OTTERS WALK
STOKE RG21 1LS

REFERENCE LIBRARY
THIS COPY IS FOR REFERENCE
IN THE LIBRARY ONLY AND IS
NOT AVAILABLE FOR HOME
READING.

RSSL.116
0118844792.

LONDON: HMSO 1992

© *Crown copyright 1992*

First published 1992

ISBN 0 11 884479 2

Bibliographical reference

PEACOCK, J D, MENDUM, J R, AND FETTES, D J. 1992. Geology of the Glen Affric district. *Memoir of the British Geological Survey*, Sheet 72E (Scotland).

Authors

J D Peacock, BSc, PhD, FRSE, MIGeol.
formerly *British Geological Survey, Edinburgh*
J R Mendum, BSc, PhD
D J Fettes, BSc, PhD
British Geological Survey, Edinburgh

Contributor

D Gould, BSc, PhD
British Geological Survey, Edinburgh

Other publications of the Survey dealing with this district and adjoining districts

BOOKS

British Regional Geology
The Northern Highlands (4th edition), 1989

Memoirs
The geology of central Ross-shire, 1913
The geology of the Kintail district, in press

Technical Reports
Metagabbros in granitic gneiss, Inverness-shire and their significance in the structural history of the Moine, 77/20

Maps

1:625 000

United Kingdom (North Sheet)
Solid geology 1979
Quaternary geology 1977
Bouguer anomaly 1981
Aeromagnetic anomaly 1972

1:250 000
Great Glen Sheet (57°N 06°W)
Solid geology 1989

1:63 360

Sheet 82 Lochcarron 1913

1:50 000

Sheet 62E   (Loch Lochy) Solid and Drift 1975
Sheet 72W   (Kintail) Solid and Drift 1985
Sheet 72E   (Glen Affric) Solid 1986

Printed in the United Kingdom for HMSO
Dd 291138   C10   12/92   20249   3396/2

Geology of the Glen Affric district

This memoir describes the geology of a remote, sparsely populated area of the Northern Highlands which has a rugged mountainous topography that includes several peaks and ridges rising over 1000 m above OD. The area lies within the Caledonian orogenic belt and is chiefly underlain by metamorphosed sedimentary rocks (Moine) in which occur a few small exposures of older metamorphic rocks correlated with the Lewisian. The metamorphic rocks are cut by several suites of minor intrusions and by the major intrusion of the Cluanie Granodiorite. Large bodies of older granitic gneiss occur within the Moine outcrop.

The Lewisian rocks consist of acid gneiss and hornblende schist which occur as small bodies in the extreme north-west of the area. They are interpreted as tectonically emplaced slices within the Moine succession.

The Moine rocks were originally deposited as sandstones, siltstones and mudstones in a shelf environment. They, with the older Lewisian rocks, were metamorphosed and intensely deformed during the Lower Palaeozoic Caledonian orogeny and by earlier episodes of folding and metamorphism. The sandstones have been converted to quartzofeldspathic rocks (psammites), the shales to mica-schists (pelites) and the siltstones to quartz-mica-schists (semipelites). Further south the Moine rocks have been subdivided into three tectonostratigraphic divisions, termed the Morar, Glenfinnan and Loch Eil divisions, and representatives of each have been recognised in the Glen Affric district. However, the complexity of folding and and sliding, lack of distinctive marker horizons has prevented the establishment of a regional stratigraphy although successions of local significance in the Glen Affric district are described.

The intense folding of the Moine rocks is the result of both pre-Caledonian and Caledonian episodes and three major tectonic events (D_1, D_2, D_3) have been recognised. Folds correlated with the D_2 event are recognised throughout the district but structures correlated with the later D_3 event are less widespread, being most common in steeply dipping rocks of the west and north.

In the Glen Affric district the Morar Division is separated from the Glenfinnan Division by the Strathconon Fault. The contact of the Glenfinnan Division with the Loch Eil Division is regarded as having been originally stratigraphic but the boundary has been modified by intense folding and ductile thrusting. Folded masses of granitic gneiss are present close to the contact between the two divisions. The granitic gneiss bodies are interpreted as at least partly tectonically emplaced pre-Caledonian granitic intrusions that have been folded and metamorphosed. Minor basic igneous bodies, now represented by hornblende-schists, amphibolites and metagabbros, were intruded into the Moine rocks before the onset of the Caledonian orogeny. During the later stages of the Caledonian orogeny the Cluanie Granodiorite and associated dykes of felsic porphyrite were intruded. The subsequent post-orogenic intrusion of microdiorites, felsites and lamprophyres continued into Lower Devonian times. Further igneous activity during the late Carboniferous to Permian is represented by the occurrence of camptonite-monchiquite dykes.

The pre-Quaternary land surface was modified by repeated glaciation throughout the Pleistocene. The valley floors are now mantled by till and hummocky moraines deposited partly during the Loch Lomond Readvance, some 11 000 to 10 000 years ago, and partly during earlier episodes of the main Late-Devensian glaciation.

HAMPSHIRE COUNTY LIBRARY

WITHDRAWN.

REFERENCE LIBRARY

THIS COPY IS FOR REFERENCE IN THE LIBRARY ONLY AND IS NOT AVAILABLE FOR HOME READING.

C002440540

Glen Affric: looking west towards Sgurr na Lapaich (1036 m), An Tudair and Mam Sodhail.
Loch Affric in the foreground with relict Caledonian forest of Scots pine and birch.

CONTENTS

FIGURES

PLATES

TABLES

PREFACE

The sparsely populated area described in this memoir is one of isolated, rugged beauty which attracts climbers, ramblers and others who share the love of the outdoors. Many of these people wish to better understand how this area has evolved geologically and this memoir will help them to do this. However, the region also has a fundamental geological significance in that it lies in the heart of the Caledonian mountain chain with a geology dominated by the intense deformation, metamorphism and igneous intrusion which produced the Caledonian orogenic belt in late Precambrian and early Palaeozoic times. This memoir describes the geology revealed by the primary survey of the area undertaken between 1964 and 1978 and makes a significant contribution to our understanding of the many phases of deformation superimposed on the sequence of ancient shelf sediments which make up the bedrock and which, with the numerous intrusions, are responsible for the very complex outcrop patterns evident on the 1:50 000 map. However, even after over 100 years of study by a host of university, Geological Survey and company geologists our understanding is still incomplete. We are not yet able to fully relate the Glen Affric geology precisely to the development of the Caledonian mountain chain as a whole.

The other major influence on the landforms of the area was the Quaternary ice ages; the repeated glaciation and deglaciation sculpted the topography to its present form and left a range of well-preserved deposits as the ice retreated. The detailed interpretation of these deposits and of the geomorphology over such a large area has added considerably to our understanding of the Quaternary geology of the Highlands. At a time when we are striving to better understand the processes involved in climatic change, the documentation of past environments in Britain and the waxing and waning of ice caps is critical to the development of climatic models for western Europe, and the British Geological Survey has an important contribution to make to our understanding of past climates.

The Glen Affric survey has shown there is some potential for sand and gravel resources among the Quaternary valley deposits, and there are small shows of base metals in fault zones on the borders of the area. But overall there is little indication of economic mineral deposits, and the most valuable natural resource known in this area is undoubtedly its unspoiled natural beauty.

Peter J Cook, DSc
Director

British Geological Survey
Keyworth
Nottingham
NG12 5GG

23 April 1991

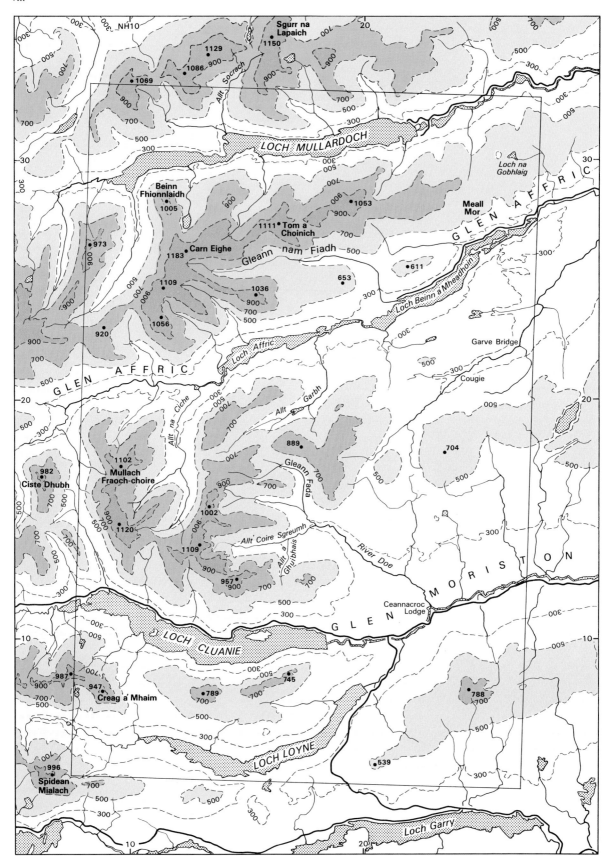

Figure 1 Topography and place names.

ONE

Introduction

The area covered by Sheet 72E—Glen Affric, is one of varied relief, and includes several peaks rising more than 1000 m above OD (Figure 1). The craggy mountains of the west pass eastwards into more rounded ridges which overlook lower hills and broad straths. Changes in vegetation and precipitation correspond roughly with this passage, the grassland of the west being replaced by heather-clad hills farther east. Records kept by the North of Scotland Hydro-Electric Board show that rainfall is some 2.5 to 3.0 m per annum on or near the high ground on the west side of the Sheet, but only about one metre on the low ground in Glen Moriston (NSHEB Annual Reports, 1956–1966).

The dissected ground around Loch Mullardoch, Glen Affric and Loch Cluanie retains traces of erosion surfaces. Godard (1965, pp. 325–326) pointed out the accord of summit levels at 970 to 1000 m above OD north of Loch Cluanie and at 890 to 990 m above OD in the broad area south-west of Loch Mullardoch. These he ascribed to his 'surface supérieure' of northern Scotland. East of the highest ground a locally prominent surface occurs at about 550 m above OD in the area north of Loch Affric and another at about 250 m on the north side of Glen Moriston adjacent to the outlet of the River Doe. The 550 m surface, together with other remnant surfaces at a slightly lower altitude about Cougie [NH 240 210], comprise the 'surface intermédiare' of Godard, and the 250 m surface can be correlated with his 'surface écossaise'.

Large-scale glacial sculpturing extensively modified the 'pre-glacial' landscape to produce the three principal east–west valleys now occupied by Loch Mullardoch, Loch Affric and Loch Cluanie, and the corries on the north and east sides of the higher hills. 'Knob and lochan' topography produced by the passage of glacier ice is well developed in the eastern half of the sheet south of Loch Beinn a'Mheadhoin. The three large lochs referred to above, together with Loch Beinn a'Mheadhoin and Loch Loyne, occupy ice-excavated rock basins.

The area lies within the region north of the Great Glen in which no detailed geological mapping had taken place prior to 1945. The region has since been completely mapped by the Geological Survey. No geological description of any kind within the area of Sheet 72E existed until the publication by Leedal (1953) of a detailed study of the Cluanie Granodiorite body together with an account of the surrounding Moine. Leedal distinguished two broad divisions in the Moine rocks of the area, which can be correlated with the Glenfinnan and Loch Eil divisions later recognised over a wider area by Johnstone et al. (1969). He also recorded a steep belt of highly inclined strata in the west and a flat belt of gently inclined beds in the east, the two belts trending north-east, roughly parallel to the Great Glen. The junction of the two belts corresponds more or less to the boundary between the Glenfinnan and Loch Eil divisions and was termed the Loch Quoich Line by Clifford (1958, p. 72). In the north of Sheet 72E, the Moine of the lower reaches of Glen Cannich and Glen Affric was mapped more recently by Tobisch (1963, 1966 and 1967). Aspects of the minor intrusions (metamorphosed and unmetamorphosed) and of the metasomatic rocks on the sheet have been studied by Dearnley (1967), Tanner and Tobisch (1972), Peacock (1977), Smith (1979), Rock (1983, 1984) and Rock et al. (1985). The Quaternary history of parts of the Glen Moriston and Glen Affric areas has been discussed by Peacock (1975) and Sissons (1977) and aspects of the geomorphology were described by Godard (1965).

The account of the geology which follows is based upon this earlier and later work and upon the systematic 1:10 560 and 1:10 000 mapping undertaken by the Geological Survey between 1964 and 1978. This mapping was undertaken mainly by Drs N G Berridge, G C Clark, D J Fettes, W G Henderson, J R Mendum, J D Peacock, N M S Rock and D I Smith, with Mr G S Johnstone as District Geologist. Small areas were surveyed by Drs A L Harris, F May, and C G Smith and by Mr G S Johnstone. Chemical analyses were carried out by Mr B H Walker. Figure 29 and the discussion of chemical analyses in chapter five are based on work by Dr A J Highton. Chapters one to three and five to nine were compiled and written by Dr Peacock. Chapter four was written by Drs Peacock and Mendum and includes an account of the metamorphism contributed by Dr Fettes. Dr D Gould contributed the section on the Cluanie Granodiorite in chapter six. The memoir was edited by Dr W Mykura, Dr D I J Mallick and Dr G C Clark.

TWO

Regional setting and summary of the geology

The Glen Affric district (Figures 2 and 3) is underlain chiefly by metamorphosed sedimentary rocks of the Moine succession, a Late Proterozoic shallow-water sedimentary sequence that was intensely deformed and metamorphosed some 450 Ma ago during the Caledonian Orogeny; it was also probably affected, in part at least, by earlier deformation events at about 750 Ma and 1000 Ma ago.

The Glen Affric area lies wholly within the Caledonian orogenic belt of the Northern Highlands, the north-west margin of which lies 20–30 km to the west and is marked by

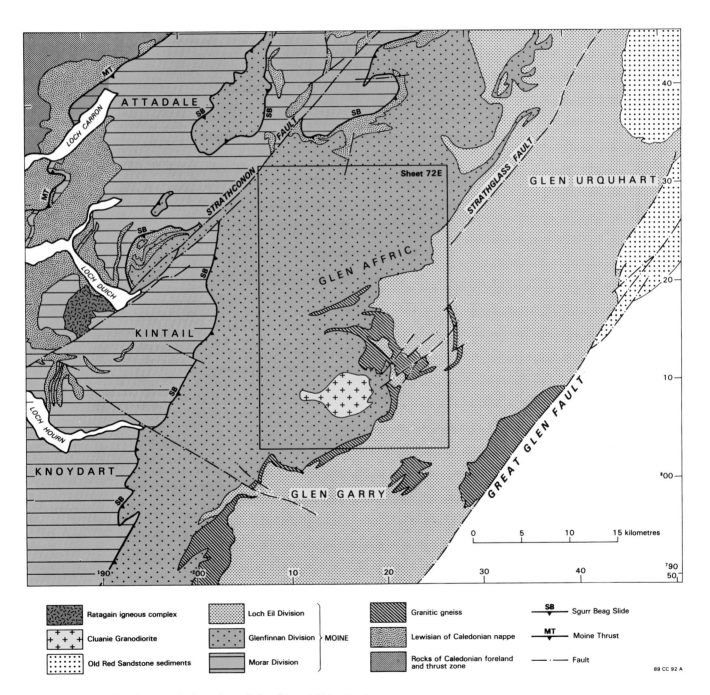

Ratagain igneous complex	Loch Eil Division		Granitic gneiss	**SB** Sgurr Beag Slide
Cluanie Granodiorite	Glenfinnan Division	MOINE	Lewisian of Caledonian nappe	**MT** Moine Thrust
Old Red Sandstone sediments	Morar Division		Rocks of Caledonian foreland and thrust zone	Fault

89 CC 92 A

Figure 2 Regional geological setting of the Glen Affric district.

the Moine Thrust Zone, a zone of west-north-west-directed thrusts along which the rocks of the orogen were emplaced on to the foreland Lewisian gneisses and Torridonian red-bed sediments. The Lewisian rocks formed the basement upon which the Moine succession was deposited and they now also outcrop east of the Moine Thrust Zone as deformed and metamorphosed slices folded and faulted into the Moine. Only one small area of such Lewisian rocks occurs on Sheet 72E, in the north-west corner, although they are widespread on adjacent sheets both to the west (72W) and north (82E).

As a consequence of the general lack of good marker horizons in the Moine succession, and the complex, multiphase deformation that the rocks have suffered, neither the stratigraphy nor the structure is yet fully understood in detail. It has been possible, however, to extend the three-fold stratigraphic–structural subdivision of the Moinian recognised further south into Sheet 72E (Johnstone et al., 1969). The lowest, Morar, Division is confined to the extreme north-west, north of the Strathconon Fault. The remainder of the sheet is divided fairly evenly between the mixed lithology Glenfinnan Division and the dominantly psammitic Loch Eil Division.

Three major tectonic events (D_1, D_2 and D_3) have affected the Moine rocks, resulting in complex interference fold patterns and with appreciable slip along some lithological boundaries. One major discordant slide has been recognised. South of Sheet 72E the Glenfinnan Division rocks mostly have steeply dipping foliations while those of the Loch Eil Division are mainly gently dipping, the boundary between the two being the narrow structural zone of the Quoich line (Clifford, 1957) which has been interpreted as the eastern limit of late Caledonian deformation of the orogen (Harris, 1983, Roberts et al., 1987). This subdivision into 'steep belt' and 'flat belt' has been extended into the Glen Affric area, but in the north of the sheet the trace of the Quoich line is less well defined.

In the Sheet 72E area the **Lewisian** rocks comprise acid gneiss and hornblende-schist and occur only as small bodies surrounded by Moine pelite in the valley of the Allt Coire Lungard (National Grid squares NH 09 30 and 09 31). These are interpreted as tectonically emplaced slices.

The **Moine** rocks were originally deposited as clastic sediments, but subsequent metamorphism (within the amphibolite facies) has converted the sandstones into granular quartzofeldspathic rocks (psammites), the shales into mica-schists (pelites) and the sandy shales or siltstones into quartz-mica-schists (semipelites). Thin beds and lenses of calcareous sediment, mainly diagenetic impregnations, are now represented by ribs and pods of calcsilicate rock. There are a few bands of highly garnetiferous rock containing hornblende which may be partly of volcanic origin.

In the Glen Affric district the Moine metasediments are grouped into three units correlated with the Morar, Glenfinnan and Loch Eil structural–stratigraphic divisions of Johnstone et al. (1969) (Figure 2). Of these, the **Morar Division**, which is entirely psammite, occupies only a very small area in the extreme north-west. The **Glenfinnan Division** consists mainly of mappable units of pelite, semipelite, psammite and quartzite, but also includes thinly banded (striped) rocks in which these lithologies are interbanded on too small a scale to be mapped separately. The **Loch Eil Division** is chiefly psammite and quartzite.

The intense folding in the Moine is correlated chiefly with the D_2 (pre-Caledonian) and D_3 (Caledonian) phases of deformation. D_2 structures occur widely, but D_3 folds are restricted mainly to the steeply dipping strata in the west and north part of the district. Cross-bedding is locally preserved in the psammites, but it has not been possible to reconstruct the original stratigraphy of the Moine except within small areas because of the complex, and not completely understood, structure. The Morar Division is separated from the Glenfinnan Division by the Strathconon Fault in this district. The contact of the Glenfinnan and Loch Eil divisions is regarded as originally stratigraphic, but modified by local sliding and intense ductile folding, particularly in the pelites and mixed lithologies.

The igneous intrusions in the district have a wide range of age and composition. The earliest, which predate D_2 and probably D_1, are the folded sheets of granitic gneiss which are present at or near the boundary between the Glenfinnan and Loch Eil divisions. They are thought to be intrusive granitic rocks which were subsequently metamorphosed, folded and possibly in part tectonically introduced into their present positions. Also predating D_2 but postdating the granitic gneiss are metamorphosed basic sills and irregular bodies now represented by hornblende-schist, amphibolite and metagabbro which are particularly common within a short distance of the boundary between the Glenfinnan and Loch Eil divisions. More than one suite of such metabasites may be present.

The Late-Caledonian intrusions include suites which were emplaced before tectonic movement had entirely ceased. The earliest are transgressive veins of pegmatite which occur throughout the area. They are thought to be chiefly of Late-Caledonian age, but a few may predate D_3. Younger than the regional pegmatites are three further suites of minor intrusions as well as one major intrusion, the Cluanie Granodiorite. The latter is formed chiefly of biotite- and hornblende-bearing granodiorite which has been subdivided largely on the basis of the presence and abundance of megacrysts of potash feldspar. The disposition of structures in the adjacent rocks suggests that the granodiorite was intruded passively by stoping rather than forceably by injection. Bodies of aplite, which were emplaced during the last intrusive phase, cut the granodiorite as well as the nearby Moine rocks. Associated with the Cluanie Granodiorite is a suite of early felsic porphyrites which were intruded before the aplites and probably also before the granodiorite proper.

The Cluanie Granodiorite and its aplites are cut by sheets of the microdiorite suite which range in composition from felsic porphyrite, through microdiorite and appinite, to ultrabasic rock. Many of these intrusions are partly schistose and were metamorphosed under greenschist facies conditions. In the southern part of the Glen Affric district there is a swarm of minor intrusions of non-foliated granodiorite, which in places forms ramifying veins. The period of intrusion of this northern part of the Glen Garry vein complex overlaps with that of the microdiorite suite. At some localities in the south-east part of the district there are breccias associated with the vein complex. They are composed almost entirely of blocks of Moine country rock, amphibolites, early felsic porphyrite, and granitic gneiss.

The latest minor intrusions associated with the Caledonian Orogeny are north-west- and west-north-west-trending dykes of felsite and lamprophyre which are probably of Lower Devonian age; these crop out chiefly in the western half of the district.

Dykes of the Permo-Carboniferous camptonite-monchiquite suite occur along the northern and south-eastern margins of the map area and a single example has been found of a dolerite dyke of possible Tertiary age.

Small bodies of sodic metasomatite with hornblende and pyroxene are known from a few localities. These occurrences are similar to others elsewhere in the western Highlands which are cut by microdiorites. Metasomatic albitite spatially associated with the Cluanie Granodiorite probably postdates the Glen Garry vein complex.

Most of the major faults in the district trend between south-east and north-east. The Strathconon Fault, in the extreme north-west, has a net sinistral movement of about 8 km on the adjoining Kintail (72W) Sheet. Galena and barytes occur in specimen quantities in an east-north-east-trending crush zone in the Allt nam Peathrain north-east of the Cluanie dam [192 107].

The Quaternary history is summarised in the geological sequence shown on the inside front cover. Till deposited by the main Late Devensian glaciation occurs in the eastern part of the district, but most of the glacial deposits can be attributed to the Loch Lomond Readvance. Terminal moraines of the latter are well developed in Glen Moriston and Glen Affric, but their limit is much less clearly marked in Glen Cannich.

THREE

Lewisian

Rocks of Lewisian aspect crop out in Coire Lungard, both in the burn and as three small pods in the crags west of the corrie lip [09 31]*. In the stream [093 312] grey feldspathic gneiss, deformed pegmatite, folded hornblende-schist and biotite gneiss are exposed over a distance of about 100 m, but the contacts with the adjacent Moine pelite are not clearly seen. The grey gneiss is richer in feldspar than the psammite in the Moine nearby, from which it is otherwise distinguished only by its platiness and its association with hornblendic rocks and biotite gneiss. One of the three pods mentioned above [0958 3059]*, which consists only of platy acid gneiss, is some 25 m across and includes allanite-bearing bands. These enclaves of Lewisian are considered to be tectonic slices, emplaced in pelitic rocks of the Glenfinnan Division at a similar structural level to the much larger mass of Lewisian rocks in the Glen Strathfarrar area within the Lochcarron (82) Sheet to the north.

* National Grid reference in square NH 09 31 is to the nearest kilometre.

FOUR

Moine

INTRODUCTION

In the Glen Affric district the original clastic sediments of the Moine succession have been metamorphosed to form quartz-feldspar-granulites (granulite in this context refers to the granular fabric of the rocks and does not imply granulite metamorphic facies), semipelitic schist and pelitic schist; these are referred to informally in this memoir as psammites, semipelites and pelites. Calcareous rocks are represented only by very subordinate calcsilicate rocks in the form of ribs and lenses. These are, however, important because of their value as indicators of metamorphic grade and as local stratigraphic markers (Johnstone et al. 1969; Tanner 1976).

The Moine rocks of western Inverness-shire have been subdivided into three structural–stratigraphic units termed the Morar, Glenfinnan and Loch Eil divisions (Johnstone et al, 1969). This classification has been extended from the type area between Fort William and Mallaig to cover much of the Moine outcrop north of the Great Glen (Tanner et al, 1970, Johnstone, 1989). A continuous stratigraphy based on way-up evidence has been recognised only in the Morar Division, the other divisions being defined chiefly on a combination of structural and lithological parameters.

The characteristics of the three divisions in the type area are summarised as follows (Johnstone et al, 1969):

Loch Eil Division

Loch Eil Psammite; variably quartzose psammitic granulite (locally a micaceous 'salt and pepper' type) with very subordinate bands of pelite and semipelite. Calcsilicate ribs and lenticles present throughout and locally abundant.

Glenfinnan Division

Glenfinnan Striped Schists; banded psammite and pelitic gneiss. Contains pods and lenses of metasedimentary amphibolite and calcsilicate granulite.

Lochailort Pelite; pelitic gneiss; metasedimentary amphibolite and calcsilicate lenses usually present.

Morar Division

Chiefly micaceous and siliceous psammites (Lower Morar Psammite and Upper Morar Psammite) with an intervening assemblage of pelite and striped rocks. Local basal pelite overlies Lewisian gneisses.

In the Glen Affric district the only representative of the Morar Division is a psammite exposed in a very small area north-west of the Strathconon Fault (Figures 2 and 3). South-east of the fault the rocks are assigned to both the Glenfinnan Division, which includes large bodies of psammite, and the Loch Eil Division. The boundary between the Glenfinnan and Loch Eil divisions is taken generally to be at the easternmost occurrences of major bands of pelite. Psammite F, north and south of Loch Cluanie (Figure 3) differs from the Loch Eil Psammite further east in that it contains bands of pelite and bodies of coarse-grained, locally pebbly psammite, especially south of the southern limit of Sheet 72E. Such bands of pebbly psammite are known to occur in the rocks of the Glenfinnan Division in the area, but not in the Loch Eil Division. Current thought would assign Psammite F to the Glenfinnan Division. The boundary of this psammite against the Loch Eil Division is partly obscured by outcrops of granitic gneiss.

Structurally the Glen Affric area has been divided into 'steep' and 'flat' belts (Leedal, 1953), a broad grouping dividing rocks in the west with a foliation that is dominantly steep or vertical from rocks in the east that are moderately inclined. The boundary between the 'steep' and 'flat' belts, the Quoich Line of Clifford (1957), is a narrow structural zone that lies near to, but not coincident with, the lithological boundary separating the Glenfinnan and Loch Eil divisions.

The structural distinction between the Glenfinnan and Loch Eil divisions is thus less clear in the Glen Affric district than it is further south, though the broad lithological contrast remains.

LITHOLOGY AND LOCAL STRATIGRAPHY

General

The Moine rocks of the Glen Affric district are characterised by the repetition of a few rock types. Marker horizons of more than local significance are absent. Although sedimentary structures giving useful 'way-up' evidence have been recorded at many localities, they are insufficient to support more than local stratigraphies where there are complex folds and ductile thrusts (slides), such as in parts of the present area. The following account gives an overall view of the lithology, followed by a brief account of the distribution of the various rock types.

Psammite (quartz-feldspar granulite)

The psammites vary from almost pure white quartzites to dark grey, highly micaceous but still granular rocks in which feldspars (plagioclase and/or potash feldspar) are equal in volume to quartz and in which mica (muscovite and/or biotite) forms about 20 per cent. Accessory garnet, zoisite, epidote, sphene, apatite and iron oxide may also be present. Where abundant, iron oxide (as magnetite) has been taken as a useful marker in psammite (Ramsay and Spring, 1962). Areas where magnetite is locally abundant are shown on Figure 3 and discussed below (p.15); many such areas in the south-west of the district are in migmatised psammites.

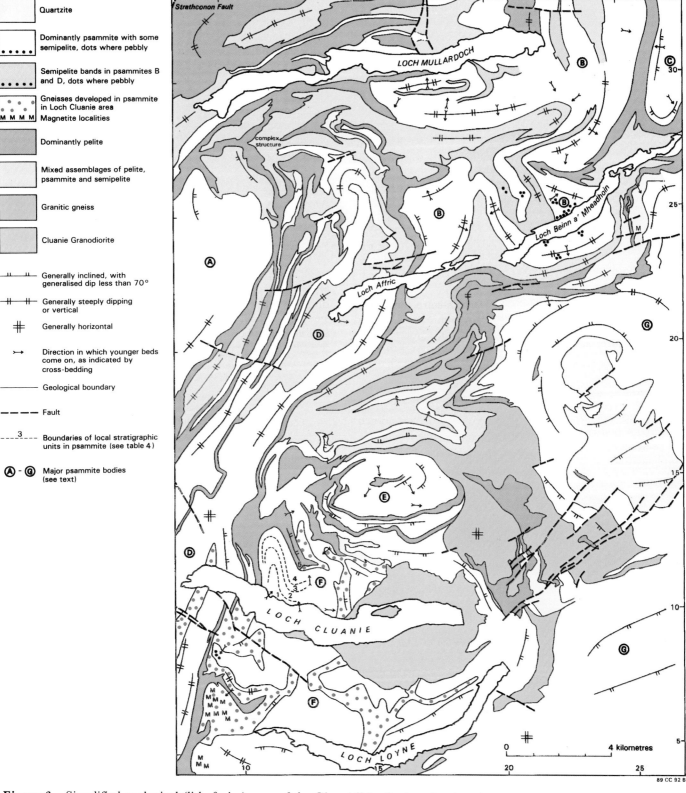

Figure 3 Simplified geological (lithofacies) map of the Glen Affric district, showing the major psammites (A–G) described in the text.

Psammite is the dominant lithology over half the map area (Figure 3); individual psammite units labelled A–G are described in detail below (p.13). It is an important component of the mixed lithologies which are described below. The most widely distributed varieties are sparsely micaceous psammite and micaceous psammite (>10 per cent mica), rock types which grade into one another and are interbanded on scales ranging from a few centimetres to many metres (Table 1). Dark, highly micaceous psammite transitional towards semipelite occurs locally, but is important only in one major body (A on the western edge of Figure 3) where it is interbanded and interlaminated with paler, less micaceous psammite. Quartzite is a subsidiary rock type in the major psammites in the western part of the district and occurs as discrete bands in mixed assemblages. Large bodies of quartzite occur within the Loch Eil Division in the east of the map area.

Clastic grains varying from coarse sand to pebble size occur at several localities, particularly near Loch Beinn a 'Mheadhoin and south of Loch Cluanie (Figure 3). It is likely that the variation in grain-size from fine to coarse in nongneissose psammite reflects to some extent the grain-size of original sand grains, now completely recrystallised. The pebbles, which are up to about 3 cm in diameter, consist entirely of quartz and potash feldspar and are usually matrix-supported. The pebbly psammite forms bands up to 3 m thick, and lines of pebbles sometimes occur in cross-sets.

In most places the mappable bands of psammite include beds of pelite and semipelite which range from thin ribs to beds a few metres thick. Micaceous partings with spacings varying from a few millimetres to a few centimetres apart occur in almost all psammites. They are probably of sedimentary origin in most cases, but may be the result of recrystallisation in others.

The very subordinate calcsilicate rocks (Figure 4 and Table 1), which are associated chiefly with psammite and semipelite, are light grey or cream coloured and are usually speckled with garnet and amphibole. They consist of quartz and varying proportions of plagioclase (andesine or bytownite), zoisite, epidote, garnet, biotite, hornblende and pyroxene. Where the calcsilicate lenses occur in psammite in which sedimentary structures are preserved, they can be seen to postdate the cross-bedding (cf. Dalziel, 1963). In these cases they are certainly replacements, probably diagenetic, rather than primary features of sedimentation. At other localities the continuous nature of the ribs and their alternation with other lithologies suggests a possible sedimentary origin (cf. Richey and Kennedy, 1939, p.29). Experience shows that calcsilicate rocks can be easily over-

Table 1
Characteristics of the major bodies of psammite indicated on Figure 3

Psammite body	Lithology	Calcsilicate ribs/lenses	Magnetite bands	Sedimentary structures	Stratigraphy (local)
A	Micaceous and highly micaceous, siliceous on E side. Bands of pelite and semipelite locally common	Locally common	Not recorded	No undoubted structures seen	—
B	Sparsely micaceous and micaceous, locally pebbly. Pelite and semipelite bands. Rare quartzite. Epidotic psammite present	Locally common	Not recorded	Common	Table 2
C	Sparsely micaceous; rare quartzite and semipelite	Rare	Not recorded	Common	—
D	Varied: see Table 3	Locally common	See Table 3	Rare	Table 3
E	Micaceous and sparsely micaceous, local pelite	Locally very abundant	Not recorded	Locally common	—
F	Micaceous and sparsely micaceous, locally pebbly. Minor quartzite, and ribs and bands of pelite and semipelite	Present locally	Present locally	Locally common	Table 4
G	Micaceous and sparsely micaceous, with quartzite. Minor pelite and semipelite	Present locally	Present locally	Cross-bedding recorded at 2 localities	—

looked in the field, particularly where clean, fresh exposure is lacking. They are thus likely to be under-represented on Figure 4. Calcsilicate rocks occur, sporadically, throughout the Glenfinnan Division, but are not abundant in much of the Loch Eil Division in this district.

Localities in which sedimentary structures are preserved in psammite, and less commonly in semipelite, occur chiefly in a belt extending from south of Loch Cluanie to the north-east part of the district, where such structures are widespread (Tobisch, 1965) and serve to define local stratigraphies. The same type and scale of structure is to be found in both the large and small bodies of psammite within the Glenfinnan

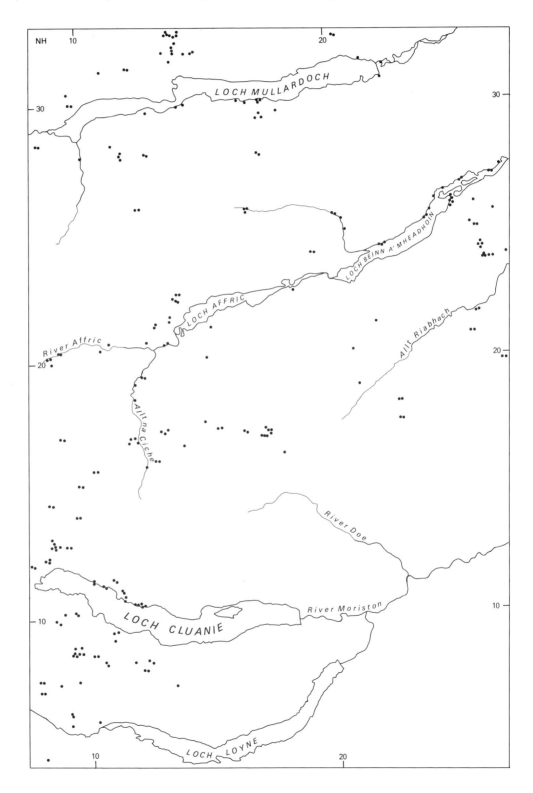

Figure 4 Distribution of calcsilicate lenticles.

Division rocks, reinforcing the impression given by the lithology as a whole that there was little diversity in sedimentary environment throughout the succession.

Cross-bedding generally occurs in sets ranging in thickness from a few centimetres to about 30 cm. It includes small-scale troughs (10 to 30 cm across and about 10 cm deep) and parallel cross-lamination. Ripple-drift lamination occurs in some finer-grained beds, including semipelite. There are some penecontemporaneous current drag folds, which deform individual units of cross-bedding, as well as slump folds and possible dewatering structures affecting several units (Plate 1). In both instances these structures are truncated by the base of the overlying bed. Good examples of most of these structures can be seen in Psammite F (Figure 3) on the north side of Loch Cluanie between localities [110 112] and [140 110] and in Psammite B on both sides of Loch Beinn a'Mheadhoinn and north of Tom a'Choinich [164 273]. At the first of these localities the psammite is locally intercalated with garnetiferous pelite suggesting that the sedimentary structures have survived amphibolite metamor-

phism with little modification. These structures are similar to those described from the Morar Division psammites by Richey and Kennedy (1939). Away from localities with well-preserved sedimentary structures, folds similar to current drag folds are locally common, but in many cases it is not possible to decide whether these are soft sediment structures or whether they are of later tectonic origin.

In some areas, notably in the southern half of the district, the psammite has been partly migmatised and shows the development of augen and lenticles of K-feldspar and quartz, typically subparallel to the bedding/foliation. The gneisses so formed occur preferentially in micaceous and highly micaceous psammites. Where these lithologies are well developed, recrystallisation has affected the whole rock and all traces of sedimentary structures have disappeared. In places concordant bands of gneiss within psammite can be mapped out, giving rise to a structural stratigraphy in areas of otherwise uniform rock type, such as between Loch Loyne and Loch Cluanie (Figure 3).

Plate 1 Cross-bedding and slump structures in psammite with micaceous laminae. Succession becomes younger towards the east (towards top of photograph). Abhainn Gleann nam Fiadh [195 269]. D2261

Bands with numerous muscovite porphyroblasts have been recorded in rock types ranging from quartzose psammite to pelite. They occur in all the larger bodies of psammite, but are particularly prominent in the northern part of Sheet 72E, in Psammite B (Figure 3). The muscovite is in some cases accompanied by quartz to form muscovite-quartz aggregates a few millimetres long which are flattened in the plane of the bedding foliation, for instance in Psammite D south of Loch Affric. Alternatively, muscovite may form discrete randomly oriented ovoidal porphyroblasts or groups of crystals, both several millimetres across, which cross-cut the lithological banding, as in Psammite B east of Loch Mullardoch. In the latter case the muscovite is in places surrounded by a bleached zone depleted in mica and rich in quartz. The porphyroblasts locally overgrow shimmer aggregate, possibly after kyanite (Tobisch, 1963). Alternatively the porphyroblasts may not be pseudomorphs after kyanite, but were formed either directly by potash metasomatism or by the recrystallisation of rock of the appropriate composition. Their presence in psammites as well as in pelites, together with the sparse occurrence of kyanite in the area, is said to favour one of these latter two possibilities (Rock, 1983).

Semipelite and pelite

With decreasing quartz, an increase in micas and plagioclase and an increase in the segregation of micas and quartzofeldspathic minerals into foliae, psammite passes into dark grey semipelite and pelite. The boundary between psammite and semipelite is taken generally to be about 20 per cent mica and that between semipelite and pelite about 40 per cent mica, though these figures must be adjusted to some extent to allow for the varying percentages of quartz and feldspars in the rock. Plagioclase (oligoclase or Na-andesine) is the dominant feldspar in both semipelite and pelite , but potash feldspar occurs locally, particularly in semipelite. Garnet, tourmaline and epidote are common accessory minerals. Hornblende has been noted in a sample of semipelite from a locality, [NH 129 200] in Gleann na Ciche south-west of the west end of Loch Affric (Rock, 1984). Kyanite has only been detected at one locality, in a semipelite within Psammite B at [1928 2956], although Tobisch (1963) recorded relict kyanite from an aluminous knot in semipelite, also within Psammite B of Figure 3 [230 303].

Within the Glen Affric district semipelite forms ribs and bands interbedded with psammite and is a component of the striped and banded rocks (see below). It also occurs interbanded with pelite. Most of the pelites and interbanded semipelites are coarse, gneissose rocks in which white quartz-plagioclase lenticles and augen are variably developed. Randomly oriented muscovite porhyroblasts occur in some bands of pelite and semipelite.

Mixed lithologies

These include psammite, semipelite and pelite which are interbanded on a scale too small to be represented separately on the 1:50 000 map. The classification also covers distinctive lithologies which, because of lack of continuity, small area and poor exposure were not mapped out. The most distinctive lithology is a rock consisting of sharply defined layers of pelite and siliceous psammite of the order of 10 cm thick, accompanied in places by subsidiary calcsilicate ribs; such rocks have distinctly striped patterns at outcrop. Calcsilicate rocks, which are characteristic of the Glenfinnan Striped Schists in their type area, are well exposed west of Loch an Goblaig [263 302] where they occur as bands in pelite. They are also found adjacent to the forestry road on the south side of Loch Affric [175 220] and in upper Gleann Fada [16 17]. A variant of the above is seen on the summit ridge of Beinn Fhionnlaidh [116 283], where thinly banded micaceous and sparsely micaceous psammite is interbanded with semipelite and pelite with local calcsilicate ribs.

Few sedimentary structures have been recorded in striped lithologies or in individual bands within the mixed lithologies where the bands are too small to be mapped. In some instances this may be due to the greater degree of strain and recrystallisation in these rocks, which are relatively weak compared to the larger bodies of psammite, but in other cases structures such as small-scale cross-bedding may never have been developed.

Big garnet rock of uncertain origin

A number of small, highly garnetiferous biotite- and hornblende-bearing bodies of uncertain affinity have been mapped at or near the contact of the Glenfinnan Division and the Loch Eil Division (Figure 21). In hand specimen they are striking rocks, with many prominent garnets up to 5 cm across which on weathered surfaces protrude above the schistose matrix. The matrix varies in appearance according to the percentage of mafic minerals, but is commonly micaceous or hornblendic and is usually studded with smaller garnets. Besides garnet, other essential minerals are quartz, biotite and plagioclase (oligoclase to labradorite). Hornblende is an essential mineral in some rocks but is rare or absent in others. Sphene and apatite are common accessory constituents together with ilmenite or magnetite. The analysis of a single hornblende-free specimen with an epidotic rather than plagioclase-rich matrix (No. 17, Table 12) falls generally within the range of Moine amphibolites (Winchester, 1976), but is relatively high in Si, Fe, P and Ba, and low in Sr.

The big garnet rocks form discrete foliated sheets a metre or less in thickness and no more than a few metres in extent. The sheets lie parallel to the bedding foliation of the enclosing quartzite, psammite, striped schist and gneissose pelite. In some cases the big garnets are developed in a pelitic or psammitic matrix in zones a few centimetres thick marginal to thin bands of hornblende-schist. No exposures have been seen in which these rocks are in contact with pre-Caledonian hornblende-schist or amphibolite of the suites described in chapter five. Big garnet rocks have not been recorded within the outcrop of the granitic gneiss (p.40), although they do occur within metasediments adjacent to it. In these cases the contacts of the big garnet rock against the gneiss may be concordant and sharp, as for instance near Ceannacroc [218 109] or concordant and more diffuse, as at a locality in Coire Dho [200 120]. At this latter locality there are transitions from big garnet rock into granitic gneiss over distances of up to a few centimetres and the gneiss near the contact is more garnetiferous than usual. Quartzofeldspathic segregations identical to those in the immediately adjacent granitic gneiss replace the matrix of the big garnet rock and isolate the large garnets.

From the above description it is clear that the rocks are characterised only by their large garnets and are otherwise mineralogically diverse, with transitional contacts into the surrounding metasediments developed in some cases. As they have not been recorded within the outcrop of the granitic gneiss and are apparently locally replaced by it they may predate the emplacement of the gneiss itself. They invite comparison in their appearance and field relationships with amphibolites interpreted as of metasedimentary origin in the Lochailort Pelite and Glenfinnan Striped Schists farther south (Johnstone et al., 1969, table II, p. 168) and with amphibolites of intrusive or extrusive origin in the Sgurr Mor Pelite (possibly Morar Division) further north (Winchester, 1976, p. 193). The different contacts with the surrounding metasediments led Peacock (1977) to suggest that they might represent either former ferruginous marly beds or, more probably, volcanic horizons, possibly water-lain tuffs. However, it is equally plausible that such contacts are the result either of metasomatism related to igneous intrusion or reactions between amphibolite and country rock during metamorphism.

DISTRIBUTION

Morar Division

The psammite on the north-west side of the Strathconon Fault in the north-west corner of the Glen Affric district (Figures 2 and 3) belongs to strata formerly termed the Siliceous Granulite Schist Series (Clifford, 1958), which has been correlated by Johnstone et al. (1969) with the Lower Morar Psammite of the Morar Division. For the most part it is a laminated, sometimes massive, sparsely micaceous to micaceous psammite with a few bands of semipelite, and at a few localities there are concordant veinlets of quartz.

Plate 2 Tightly folded lenticular band of highly garnetiferous amphibolite in coarse semipelite and pegmatite bands and lenses. Large garnets are developed near the margins of the amphibolite. South shore of Loch Mullardoch [178 302]. D3117.

Glenfinnan and Loch Eil divisions

The remaining Moine rocks of the district can be classed with the Glenfinnan and Loch Eil divisions and are taken together for the purposes of the following account.

PELITE AND MIXED LITHOLOGIES

The largest area of pelite lies north-west of Loch Mullardoch (Figure 3). Just north of the map area this pelite is in contact with a large body of Lewisian rocks on the northern flank of An Riabhachan [123 337]. To the west on the adjoining Kintail (72W) Sheet it is cut out against the Strathconon Fault. Like most pelites in the district it contains subsidiary bands and infolds of mixed rocks, including striped schists with thin calcsilicate ribs.

Further south the pelite and mixed lithologies are subsidiary to large infolds of psammite (A to E on Figure 3). The pelite separating Psammites D and F can be followed from the south-west corner of the map northwards for some 25 km to the north side of Gleann nam Fiadh [19 26]. A notable feature is the expansion of this pelite and its accompanying mixed assemblages from a band some 50 m across in National Grid square [11 16] to an outcrop more than 1.0 km across some 3 km to the south [10 14].

PSAMMITE

Psammite A (Figure 3 and Table 1)

This psammite body differs from others in Table 1 in its unusually high proportion of highly micaceous psammite and semipelite, which is locally gneissose; it has been mapped as a mixed assemblage of psammite and semipelite. It also includes bands with augened muscovite porphyroblasts. 'Wispy' folds of possibly sedimentary origin have been recorded, but no undoubted cross-bedding has been seen. Southwards on the neighbouring Kintail Sheet, it passes into the Easter Glen Quoich Psammite (Psammite D of Figure 5) across the summit of Ciste Dhubh [063 166]. This passage denotes a facies change to somewhat more quartzose and less micaceous rocks in southerly and easterly directions. The micaceous lithologies in Psammite A are identical to those developed in Psammite D in Gleann na Ciche, west of Glen Affric.

Psammite B (Figure 3, Tables 1 and 2)

Psammite B crops out northwards from Loch Affric and Loch Beinn a'Mheadhoin. A local stratigraphy can be made out in part of this psammite, which is locally pebbly (Table 2) with a few bands of semipelite in some places. Away from the Loch Beinn a'Mheadhoin area there is a lateral transition to rocks in which pebbles are sparse or absent. Epidotic psammites with abundant heavy minerals (apatite and sphene) concentrated in the micaceous laminae have been recorded in Fraoch-choire [210 290] and are common east of Loch Mullardoch.

Psammite C (Figure 3, Table 1)

This psammite, which occurs only in the extreme north-east of the map area, is lithologically similar to parts of Psammite B, with which it was correlated by Tobisch (1963). In contrast to Psammite B, however, epidotic psammite has not been recorded. The sedimentary structures, of which parallel

Table 2 Stratigraphy of Psammite B in the neighbourhood of Loch Beinn a'Mheadhoin

		Thickness (metres)
(Pelite and/or striped and banded assemblages. No sedimentary structures)		
3	Thickly to thinly banded micaceous to sparsely micaceous psammite with subsidiary bands of semipelite and pelite. Bands of quartzite occur locally and calcsilicate ribs are very common (about 12 per metre) in some places. Coarse-grained psammite present in layers up to 3 m thick. Layers with pebbles up to 15 mm in diameter present. Sedimentary structures common	700–1300
2	Semipelite with thin to thick bands of micaceous and sparsely micaceous psammite and pelite. Calcsilicate ribs locally numerous (2 to 3 per metre). Pebbles present in some psammite bands. More micaceous rocks locally gneissose. Local sedimentary structures	200–1000
1	Sparsely micaceous, coarse-grained psammite with about 25% feldspar (dominantly K-feldspar) accompanied by epidote. Bands of small pebbles. Concordance quartzofeldspathic lenticles impart a gneissose appearance. Local sedimentary structures	<300
(Base not seen)		

cross-lamination is the most frequent type (Tobisch, 1965), are like those seen in Psammite B

Psammite D (Figure 3, Tables 1 and 3)

This psammite body is the northward continuation of the Easter Glen Quoich Psammite (Geological Survey of Great Britain, 1965) extending north-north-east from the western end of Loch Cluanie to Loch Affric and Gleann nam Fiadh [15 25]. The lithological subdivisions in the line of section (Figure 5 and Table 3) can be followed northwards as far as the west end of Loch Cluanie. Near the west end of Loch Affric, Unit 8 (Table 3) passes northwards into a broad outcrop of interbanded micaceous psammite and semipelite with subsidiary pelite, and locally abundant calcsilicate ribs. North of Loch Affric, in Gleann nam Fiadh, there is a considerable development of quartzite (Figure 3) at the margin of Psammite D.

Sedimentary structures, including cross-bedding and anastomosing micaceous laminae (possible flaser bedding), have been seen at several localities around the west end of Loch Affric. They are too sparse, however, in an area of such structural complexity, to give sufficiently reliable 'way-up' evidence to define a stratigraphy.

The lithological succession in Table 3 may well involve repetition as a major tight antiform has been mapped in units 1 to 7 north of Loch Cluanie (Figure 3). The Easter Glen Quoich Psammite is folded in the Loch Quoich area by a major southward-closing fold, the Gleouraich Synform, (Figure 5) which, when traced further south, becomes the D3

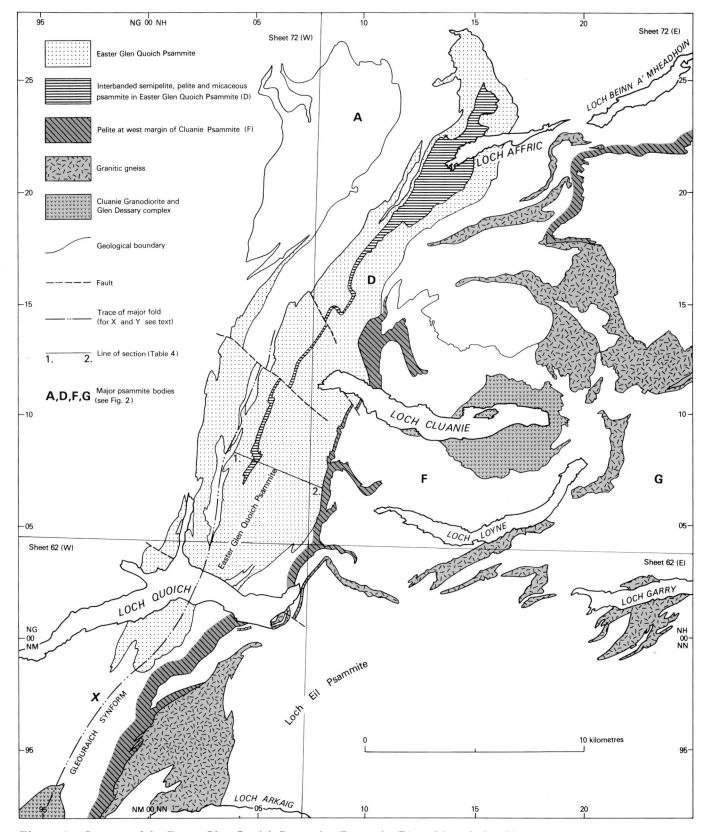

Figure 5 Outcrop of the Easter Glen Quoich Psammite (Psammite D) and its relationship to other geological features.

synform which controls the outcrop pattern of the Glen Dessary Syenite (Roberts et al. 1984). The west limb of this fold is notably attenuated compared with the east limb (Figure 5) and also contains pelite bands not found on the east limb. Hence the succession within the psammite is either structural (folds and/or slides) or there are rapid facies variations along and across the strike. A narrow band of psammite connects the west limb of the fold with Psammite A (Figure 5) in the Kintail district (Sheet 72W).

Table 3 Stratigraphy of Psammite D (Easter Glen Quoich Psammite) in Easter Glen Quoich (Figures 3 and 5)

		Thickness (metres)
(Structural top of succession to west)		
11	Laminated micaceous and sparsely micaceous psammite with quartzite bands up to 7 m thick. Rare bands of pelite	213
10	Semipelite and pelite with bands of quartzite and psammite	67
9	Laminated micaceous and sparsely micaceous psammite with thin (1 m) bands of quartzite.	460
8	(c) Striped garnetiferous pelite	60
	(b) Banded micaceous and sparsely micaceous psammite with subsidiary bands of semipelite	90
	(a) Striped garnetiferous pelite	45
7	Laminated micaceous and sparsely micaceous psammite with ribs and bands of quartzite up to 2.5 m thick. A few ribs of pelite and semipelite. Local calcsilicate nodules. Muscovite porphyroblasts in semipelite bands towards base	490
6	Quartzite with subsidiary bands of sparsely micaceous and micaceous psammite and semipelite	107
5	Laminated micaceous and sparsely micaceous psammite with subordinate ribs and bands of semipelite locally up to 20 m thick. Calcsilicate lenses towards base	790
4	Coarse, dominantly sparsely micaceous psammite with concordant lenticles of pink pegmatite up to 5 cm thick. ?Calcsilicate nodules	340
3	Semipelite and gneissose pelite with subsidiary bands of laminated micaceous and sparsely micaceous psammite	120
2	Sparsely micaceous and micaceous psammite with numerous bands of pelite and semipelite up to 2 m thick. Quartz-feldspar augen and muscovite porphyroblasts developed in some bands	340
1	Sparsely micaceous psammite and quartzite with semipelite bands. Magnetite streaks occur about 150 m from the top of the unit	1710
(Structural base of succession to east)		

Psammite E (Figure 3, Table 1)

This psammite crops out in the core of an antiformal structure which plunges to west and east to form a dome north of Loch Cluanie. It is notable locally for the high proportion of calcsilicate ribs as exemplified at one locality [140 141], where they form almost half of the rock in exposures of several metres across. Water expulsion structures resembling clastic dykes occur close to the junction of the Allt a'Ghuibhais with the Allt Coire Sgreumh [167 142].

Psammite F (Figure 3, Tables 1 and 4)

Psammite F occurs to the north and south of Loch Cluanie. The western part of the psammite, with its local calcsilicate lenticles, magnetite laminae and sedimentary structures is lithologically more diverse than the eastern part. A local stratigraphy can be made out on the north side of Loch Cluanie (Table 4) in which the strata young towards the major pelite on Meall Beac [12 11], but at other localities, for example at [122 128], the psammite seems to young away from this pelite (Figure 3, see also p.16). To the south Psammite F is continuous with rocks formerly assigned to the Loch Eil Division (Sheet 62E) but is now distinguished from the rocks of that division by the occurrence of pebbly horizons.

Psammite G (Figure 3, Table 1)

This psammite, the northward continuation of the Loch Eil Psammite (Johnstone et al., 1969) underlies a large area on the eastern side of the Glen Affric district and extends to the adjacent Invermoriston district (Sheet 73W). No firm stratigraphy, supported by younging directions, has been established. However, east of Loch Beinn a'Mheadhoin in roadside exposures adjacent to An Leth-allt [257 237] to [270 240], the rocks at its western margin are quartzitic and are followed eastwards possibly in stratigraphical order by the more typical micaceous and sparsely micaceous psammites

Table 4 Stratigraphy of Psammite F north of Loch Cluanie

		Thickness (metres)
(Youngest)		
(Pelite. No sedimentary structures)		
4	Quartzose psammite interbanded with micaceous and sparsely micaceous psammite. Sedimentary structures present locally	150 – 750
3	Gneissose psammite and pelite with quartzo-feldspathic augen. No sedimentary structures recorded	100 – 400
2	As 4	100 – 200
1	Interbanded locally gritty micaceous and siliceous psammite with bands of semipelite and common calcsilicate ribs and lenses. Sedimentary structures common	100
(Base not seen)		

For location of units see Figure 3

with quartzite and minor pelite and semipelite (Table 1) accompanied by numerous calcsilicate ribs. The state of preservation of these lithologies (for instance absence of gneisses), together with the presence of cross-cutting meta-gabbros, suggests that the apparent paucity of cross-bedding may be an original feature and that sedimentary structures are restricted to simple bedding/foliation surfaces in this part of the Loch Eil Psammite.

STRUCTURE

General

Since the fieldwork by BGS was concluded in the Glen Affric area, detailed studies have been published on the structure of areas to the south (Roberts and Harris, 1983, Roberts et al., 1984). Some of the structures described in these papers can be traced into the Sheet 72E area and used as a basis for comparison and correlation. Farther north, the structural history of the ground between Loch Monar and the eastern part of Loch Mullardoch has been summarised by Tobisch et al. (1970).

As already noted, the Moine rocks in the area fall broadly into a 'steep' belt, in which the foliation and the axial planes of folds are more or less vertical, and a 'flat' belt in which the foliation and many of the axial planes are gently dipping (Figure 3). The district has been divided into 6 sub-areas (Figure 7) to facilitate structural description. These are, as far as possible, homogeneous. Sub-areas D, E and F cover structures of the 'steep' belt, sub-area B includes structures of the 'flat' belt and sub-areas A and C extend over the narrow transition zone between the 'steep' and 'flat' belts. Three major tectonic events termed D1 to D3 can be recognised in the southern part of the district (see Roberts et al., 1984) and it is suggested that these can be correlated to a greater or lesser extent over most of the sheet. The terms 'F2', and 'F3' etc are used informally for systems of folds within each sub-area without necessarily implying correlations with the major events D2 and D3. In general the folds can be divided into 'early' or 'late', the former being chiefly of D2 age and the latter of D3 or later.

Sub-area A

This sub-area, which is characterised by late north-trending folds, lies between the 'flat' belt to the east and the steeply dipping Easter Glen Quoich Psammite (Psammite D) to the west (Figures, 5, 6 and 7). Fold amplitudes decrease considerably towards the north.

EARLY FOLDS

The earliest structure in sub-area A is a bedding foliation which is folded by early tight to isoclinal folds which are in turn refolded by the major 'F3' folds. The early minor folds ('F2') also locally fold bands of hornblende-schist and possible early quartzofeldspathic lenticles in pelite. The structural interpretation of the north–south-trending pelite immediately west of Psammite F (Figure 8) is of fundamental importance in understanding the overall structure and is therefore described in some detail.

In the south, the Allt Mialach tongue of pelite within Psammite F has been interpreted, on the basis of sedimentary structures and structural analysis, to be a fold of D2 age, refolded by the open D3 Beinn Beag synform (see also Roberts and Harris, 1983). The psammite on the south side of the tongue of pelite is pebbly and gritty, whereas that on the north side is fine to coarse grained with streaks of heavy minerals. The younging is consistently away from the pelite.

Farther north, the Allt Giubhais tongue of pelite is flanked on its south side by psammite with heavy mineral streaks (Figure 8), but these are absent from the psammite on its northern flank. Cross-bedding indicating that the psammite faces away from the pelite has been found only at one locality [094 072], the closure of the pelite being unexposed. However, there are several small interdigitations of pelite and psammite (Figure 8, and Sheet 72E). These tight to isoclinal fold closures are accompanied by an axial-plane schistosity. These are probably folds of 'F2' age and suggest that the Allt Giubhais tongue is itself a closure of this generation. The most easterly interdigitation (B, Figure 8), is refolded by minor 'F3' folds. The gneissose foliation defined by quartzo-feldspathic lenticles in the psammite adjacent to the Allt Giubhais tongue also appears to be folded by the major ('F2') closure, and the distribution of gneissose and non-gneissose psammite about the fold [09 08] indicates the presence of a complementary westward-closing fold a short distance to the north (Figure 8). Cross-bedding evidence on the north limb of the complementary fold, indicates the beds young to the south [09 09].

On the north side of Loch Cluanie the succession in Psammite F youngs north towards the Allt Coire Lair pelitic tongue. In the vicinity of the closure of the tongue, cross-bedding also indicates that the psammite youngs towards the pelite. This closure, like the other two, is taken to be an 'F2' fold core. North-east of the pelite tongue, however, cross-bedding in the psammite, at two localities [118 144] and [122 138] not far from the boundary with the pelite, shows that the strata young away from the pelite. The opposed younging directions on the northern side of the fold suggest either the 'F2' fold limb, or part of it, has been removed by sliding, or the 'F2' folds are refolding earlier structures. Cross-bedding indicates that the psammite youngs away from the pelite at most localities and at only one locality it is likely that the pelite is older than the Psammite F.

West of the Allt Coire Lair tongue, the interdigitations of psammite and pelite (squares [10 11 and 10 12]) also likely result from early 'F2' folding and this may partly explain the disposition of the quartzites, pelites and psammites east of the tongue in the area around Sgurr nan Conbhairean [130 138] in sub-area C.

LATE FOLDS

A number of major late folds lie within the west part of Psammite F and their axial traces trend roughly north–south (Figure 8 and 9). They vary in style from open to almost isoclinal (Table 5), and their north-trending axial planes are vertical or steeply dipping. An axial plane crenulation cleavage is developed in pelitic bands. The early ('F2') axial-plane schistosity in the pelite is folded by these late folds.

Figure 6 Major folds in the Glen Affric district.

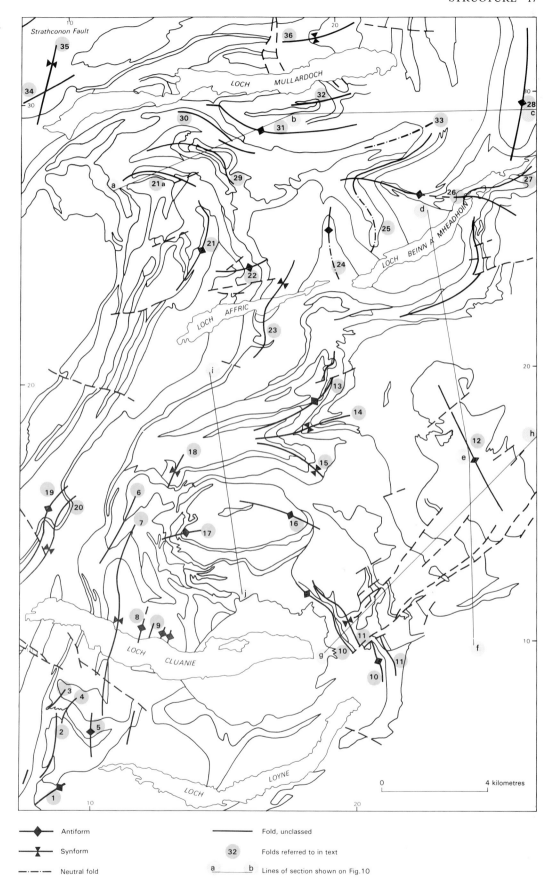

Antiform

Synform

Neutral fold

Fold, unclassed

32   Folds referred to in text

a _____ b   Lines of section shown on Fig.10

Figure 7
Structural sub-areas
described in the text.

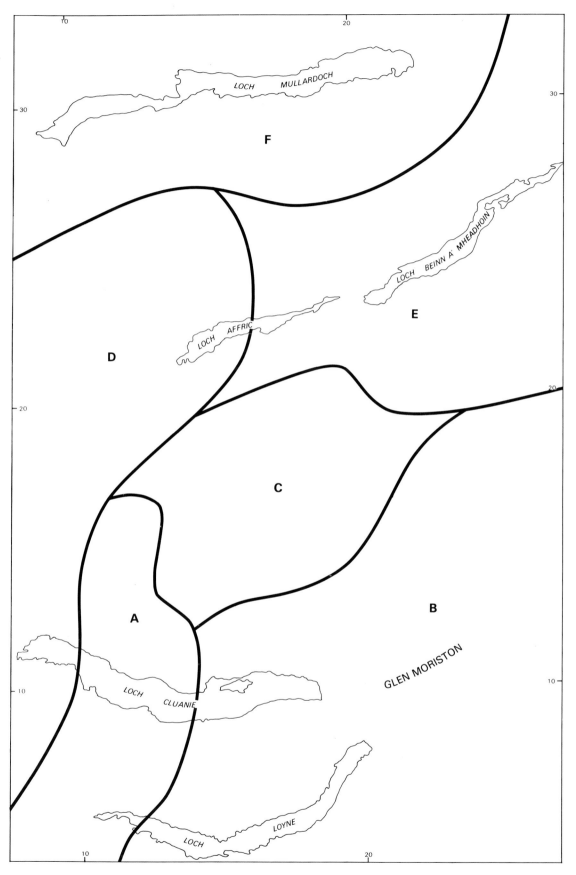

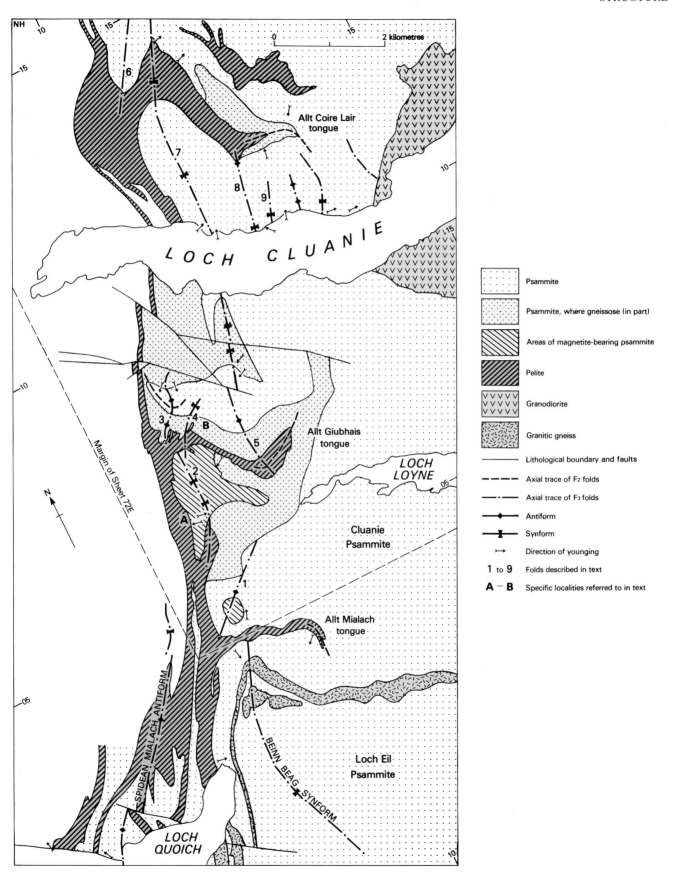

Figure 8   Structural relationships of the rocks of sub-area A to those between the sheet boundary and Loch Quoich.

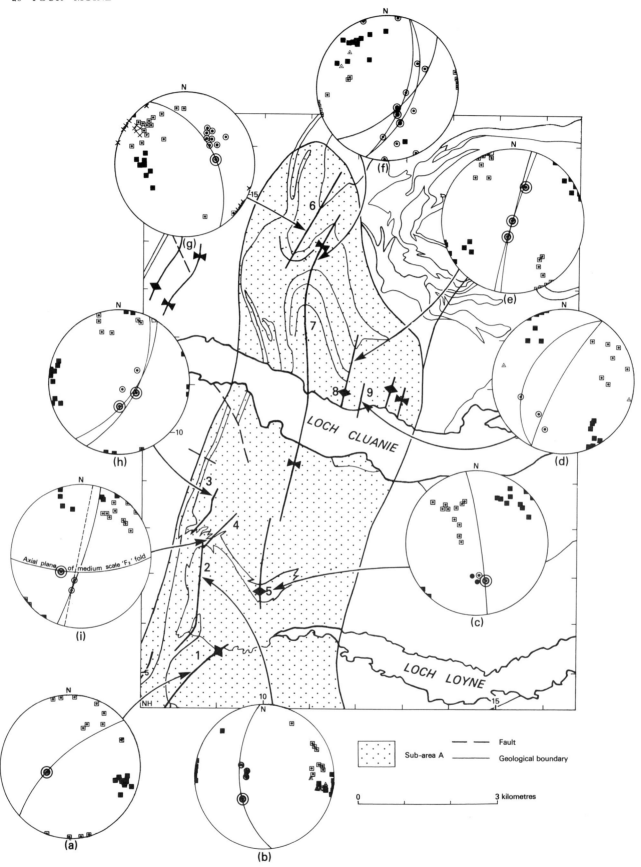

Figure 9 Stereographic projections of folds 1 to 9 of Figure 8, sub-area A. For key to stereoplots see Figure 16, p.28.

To the west, the 'F3' structures become generally tighter and the 'F2' structures are rotated into parallelism so that 'F2' and 'F3' fold structures become more difficult to distinguish. At one locality (A on Figure 8) the two sets of minor folds can be separated on the basis of the development of a crenulation cleavage in the pelitic 'F3' fold hinges and an axial-plane schistosity in the 'F2' folds. Such a distinction is not always possible as an axial-plane schistosity is also developed in tight 'F3' folds, notably in pelitic hinge zones.

The west limb of fold 6 (Figures 8, 9) is affected by later minor- to medium-scale folds with vertical north-east-trending axial planes (Figure 8) and an associated crenulation cleavage. No major folds of this latest generation are known with certainty, though they may be responsible for the broad swings of strike of the lithological banding and the axial plane of fold 7 in the same area (Figure 8).

Table 5 Summary of data on 'F3' folds in sub-area A (see Figure 9)

Fold	Style	Axial-plane strike	dip	Plunge	Remarks
1 Antiform	Open	NE	Steep to S	Steep to WSW	Early minor folds within lithological banding
2 Synform	Tight, almost isoclinal	N	Steep to W	Steep to SW	Local crenulation cleavage to axial-plane schistosity Refolds medium-scale tight to isoclinal folds with axial-plane schistosity
3 Antiform	Open to tight	NNE	Steep to ESE	Steep to between S and E	Minor folds plunge between S and NE
4 Synform	Open	NNE	Vertical	Steep to SSW	Refolds a medium-scale isoclinal fold which has an axial-plane schistosity inclined steeply S and plunging to the SW
5 Antiform	Open	N	Vertical	Steep to S	Refolds Allt Giubhais tongue. Associated with crenulation cleavage in pelite. Local lineation of feldspar augen
6 Sideways	Open to tight	NNW	50°–60° to ENE	30°–40° to E	Folds schistosity/lithological banding in pelite. W limb crossed by later NE-trending minor folds plunging NE associated with crenulation cleavage
7 Synform	Tight	Between NNW and NNE	Vertical or steep to E	About 50°S	Crenulation cleavage in pelite. Local rodding in quartz lenticles. Refolds Allt Coire Lair tongue and folds schistosity in pelite. Refolds tight minor folds in psammite
8 Antiform/ synform	Open	NNE	Vertical	About 75°S (in N) through vertical to 50°N (in S)	Folds schistosity/lithological banding in pelite
9 Synform	Open to tight	NNE	Steep to WNW	35° to SW	Crenulation cleavage in pelite, locally almost axial-plane schistosity

CORRELATION

The major late folds described above are similar in attitude
and style to the Beinn Beag synform and Spidean Mialach
antiform to the south (Figure 8) which are interpreted by
Roberts and Harris (1983) as being of D3 age. The early
'F2' folds of the sub-area can be correlated with the D2 structures of Roberts and Harris (1983) on the basis of style and
structural geometry. No D1 minor folds have been identified
with certainty in sub-area A.

Sub-area B

This sub-area occupies most of the south-east quarter of the
Glen Affric district (Figure 7), and includes psammite units
which generally dip gently to the south-east (Figure 10e–f).
They constitute part of the 'flat' belt and are in part contiguous with the Loch Eil Psammite of the neighbouring Invermoriston (73W) and Loch Lochy (62E) sheets. In the Glen
Affric district, however, the areas of Loch Eil Psammite are
separated by rocks assigned to the Glenfinnan Division and

by granitic gneiss in Glen Moriston (Figure 1). Farther
south, granitic gneiss, possibly lying along a slide, separates
the Loch Eil Psammite from Psammite F (Figure 3).

EARLY FOLDS

Early ('F2') recumbent minor folds are present, the axial
planes of which are parallel or subparallel to the foliation
(Figure 11 (a) and (c)). The folds generally vary from tight to
isoclinal but may be locally open in style. They are accompanied by an axial-plane schistosity, and a strong mineral
lineation is present in the bedding-plane foliation and the
quartzofeldspathic lenticles in the psammite. Important relationships are also seen in the River Doe section (Peacock,
1977), where the 'F2' minor folds also fold an earlier
schistosity or foliation in both granitic gneiss and hornblende-schist as well as the compositional banding in the latter (Figure 12). A major early recumbent isoclinal fold closure is present 2 km north of the east end of Loch Cluanie in
the vicinity of the late folds 10 and 11 (Figures 6 and 10g–h).

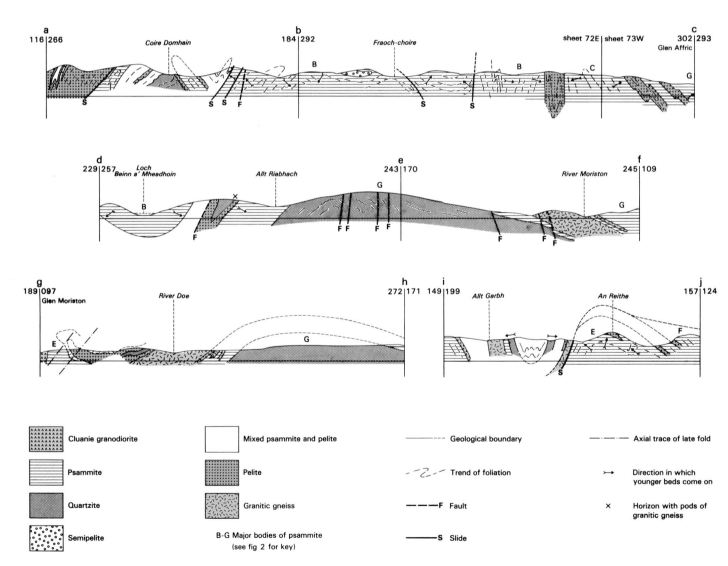

Figure 10   Diagrammatic sections across the Glen Affric district.

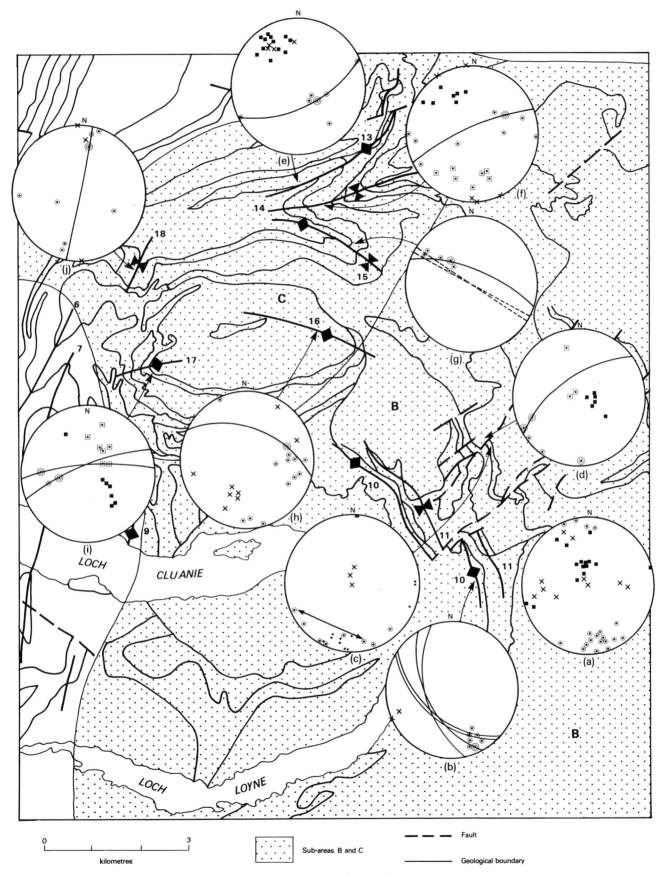

Figure 11 Stereographic projections of major and minor folds, sub-areas B and C. For key to stereoplots see Figure 16, p.28.

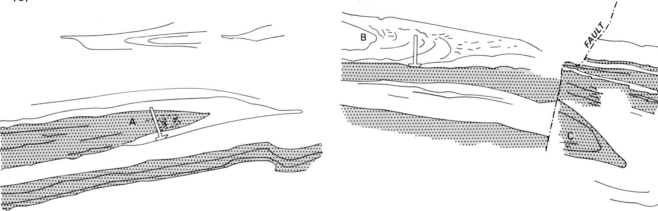

Figure 12 Sketches of early minor folds in the lower River Doe area:
(a) Recumbent tight folds in hornblende-schist and granitic gneiss.
(b) Recumbent tight to open minor folds in hornblende-schist and psammite.
(c) Recumbent tight to isoclinal minor folds in granitic gneiss and hornblende-schist.
(d) Continuation of folds in (c) to right.

LATE FOLDS

The early folds are refolded by at least two sets of late folds. East of Loch Cluanie, folds 10 and 11 (Figure 6) comprise an antiform and complementary synform which plunge gently south-south-east and have axial planes that dip about 45°SW (Figures 10g–h and 11b). Minor folds of similar trend and style, locally with an associated crenulation cleavage, occur widely east, north-east and south-east of Loch Cluanie (Figure 11a). A few of these belong to a fold set with axial planes inclined steeply to the east. They are co-axial with the minor folds associated with structures 10 and 11 and with the early folds described above.

In the bed of the River Doe (Peacock, 1977) several sub-horizontal plunging very open minor to intermediate scale folds have axial traces trending between north and north-east, (Figure 11d). They fold early minor folds, but no relationship has been established with other late structures. No accompanying cleavage or lineation were noted.

The major dome (12 on Figure 6) is a very open structure (Figure 10d–f and g–h). It is later than early tight to isoclinal recumbent folds of the style described above, but its relationship to other late folds is unknown.

The south-west boundary of the major sheet of granitic gneiss which crops out north of Loch Cluanie cuts across the stratigraphy of the Moine country rocks (Figure 3). Further north and west it coincides with a structural discontinuity (Figure 10i–j). Although there is no supporting evidence indicating high strain, such as the platy zones reported by Rathbone and Harris (1979) elsewhere in the Moine, nevertheless there is a reversal in younging across this boundary and it is suggested that this discontinuity is a slide.

CORRELATION

Early minor folds of the style noted in sub-area B also occur in sub-areas A and C, where they are regarded as D2 in age (Roberts and Harris, 1983). The presence of the major re-

Figure 13 Sketches of minor folds associated with fold 16 of Figure 11. Psammite stippled, interbanded psammite and pelite blank. East bank of Allt Coire Sgreumh [176 145].

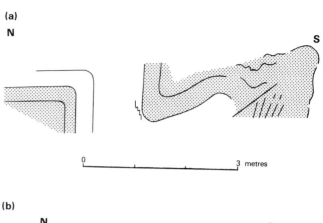

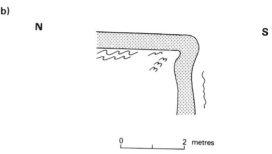

cumbent isoclinal fold north-east of Loch Cluanie (Figure 10g–h), together with the widespread tight to isoclinal recumbent minor folding in the sub-area, suggest that other major D2 folds may be present in the 'flat' belt. The folds 10 and 11 (Figure 6) postdate D2 minor folds and, as they are developed on the flank of anticline 16 in sub-area C (p.26), may postdate it. The inferred slide discussed above seems to predate folds 11 and 12 and be synchronous with or postdate D2 folding. It is interpreted either as late D2 structure or, alternatively, it may be associated with the post-D2 major folds in sub-area C.

Sub-area C

This sub-area contains the major folds 13 to 18 which generally trend west to east (Figures 6 and 11). They lie between the 'flat' belt and the north-east-trending folds of the 'steep' belt to the west and north. The major characteristics of the individual folds are given in Table 6.

Table 6
Summary of data on major folds in sub-area C (see Figure 11)

Fold		Style	Axial-plane strike	dip	Plunge	Remarks
13	Sideways closing	Tight to isoclinal	ENE	Steep to SSE	SSE	Closes to ENE. Axial plane schistosity. Tight to isoclinal minor folds with attenuated limbs. Rodding in quartz lenticles
14	Synform	Open to tight	ENE	Steep to NNW	Gentle to ENE	Refolds limb of 13 and isoclinal folds. Associated crenulation cleavage and minor folds
15	Synform to N	Open to tight	SE	Steep to NE	Moderate to NW	Local crenulation cleavage. Open to tight minor folds refold tight folds (early minor folds?)
16	Antiform	Open	E	Moderate to N	Gentle to E	Crenulation cleavage. Minor folds tight to open, fold early minor folds, but folded by open minor folds plunging gently SSE. Latter with axial planes and crenulation cleavage dipping SW
17	Antiform	Open	E	Steep N	Moderate to W	Folds early recumbent minor folds with axial-plane cleavage
18	Synform	Open	NNE	Vertical	Gentle to N	Folds early minor folds with axial-plane cleavage. Developed on S limb of fold 14

EARLY FOLDS

Fold 13 is a major tight to isoclinal early closure (Figures 6 and 11e) the geometry of which is described in Table 6. Minor tight recumbent folds similar in style to those on the limbs of fold 13 are widespread throughout the sub-area, where they fold bedding/foliation in psammite.

LATE FOLDS

Folds 14 to 17 (Figure 11) are similar in style to each other and are characterised by axial planes dipping moderately to steeply north (Table 6). The folds 14 and 15, which plunge in opposite directions (Figure 11f, g), are developed on the southern, upper limb of the early closure 13. Fold 14 is an easterly plunging synform; fold 15 is the parallel antiform in the west but its plunge steepens to the south-east until it becomes a north-westerly plunging synform. To the west, however, in the vicinity of the section shown in Figure 10i–j only a single synform is present, which is a continuation of fold 14. Well-developed cross-bedding in units c. 30 cm thick in a quartzite band suggests that the bedding is inverted in the synform.

Folds 16 and 17 are broadly anticlinal and in part define a dome (Figures 6 and 11). Minor folds developed on the limbs of fold 16 refold earlier minor structures but are themselves locally refolded by late open minor folds which plunge gently south-south-east (Table 6). The facing relationships shown in Figure 10i–j strongly imply that there is a structural break between fold 14 and fold 16/17, at or close to the band of granitic gneiss. As discussed above (p.24), this break is probably a slide.

The synform 18 (Figure 11j) is apparently developed on the south limb of the westward continuation of synform 14 and trends at a high angle to the sparse minor folds which may be associated with the synform.

CORRELATION

Fold 13 and the early minor folds of sub-area C are similar in style to the early folds (D2) of sub-areas A and B and have been correlated with them. Folds 14 to 17 resemble the D3 folds of sub-area A, but differ markedly in trend. Fold 18 continues the trend of the folds correlated with D3 in sub-area A and is probably of this D3 group. Its relationship to the sparse minor folds apparently associated with fold 14 suggests that the latter could be earlier (pre-D3), but the relationship requires confirmation. The late south-south-east-plunging minor folds which postdate structures associated with fold 16 are compatible with the style of the major fold pair 10 and 11 nearby in sub-area B and support the view that the latter are also later than fold 16.

Sub-area D

This sub-area is defined by the outcrop of Psammites A and D on Figures 3 and 5 together with the intervening strata. The axis of the major D3 Gleouraich Synform (Roberts et al., 1984), (X on Figure 5) probably extends into the Glen Affric district between Psammites A and D, but its northward continuation in the map area is not clearly seen.

EARLY FOLDS

The earliest folds ('F2') in the sub-area are tight to isoclinal minor folds with attenuated limbs and a strong axial-plane schistosity well developed in the pelitic units. These structures fold the bedding/foliation in psammite and are in turn folded by late 'F3' minor folds described below. A major early fold can be seen in the crag south-south-east of Mullach Fraoch Choire (Figure 14) and a zone of intense minor folding may define another 'F2' hinge zone just west of Loch Affric (Figure 15). In the latter area there is a strong rodding lineation (bedding foliation/axial-plane schistosity intersection) related to 'F2' which plunges steeply both to the north-east and west-south-west, possibly as a result of 'F3' folding.

LATE FOLDS

Folds 19 and 20 ('F3') are a complementary antiform and synform and have eastward-dipping axial planes (Figures 6 and 15, and Table 7). The steep west limb of the antiform contains a major early fold hinge (Figure 14). The late minor folds associated with the fold pair can be traced north-eastwards into Gleann na Ciche, where the dip of their axial planes steepens to subvertical. Muscovite porphyroblasts are recrystallised in a tight crenulation cleavage to lie parallel to the axial planes of these minor folds.

A stereographic plot of the late minor folds and foliation immediately south-west of Loch Affric suggest the presence of a very tight antiform with a south-south-west-plunging axis (Figure 16b) between the stream junction [120 190] and the River Affric. No major fold has been defined in the field, however, owing to lack of stratigraphical marker horizons and the complex minor folding. The late 'F3' minor folds refold early isoclinal folds with their attendant penetrative axial-plane schistosity. They are locally crossed by a vertical east–west-trending coarse crenulation cleavage associated with post-F3 open minor folds which are a continuation of structures associated with fold 22 in sub-area D.

On the north side of Glen Affric, west of Loch Affric, the 'F3' minor folds can be considered in 3 zones (Figures 15 and 16d). In the south-east, adjacent to Loch Affric, the 'F3' minor folds, which face upwards according to cross-bedding evidence, plunge chiefly to the north-east. Their vergence indicates an antiformal closure to the north-west. In a central complex zone, which coincides with numerous 'F2' minor folds, their plunge is variable and varies through the vertical from north-east to south-west and the 'F3' minor folds apparently locally face downwards. To the north-west, the 'F3' minor folds plunge consistently south-south-west in accordance with the attitude of similar folds along strike to the south of the River Affric. Their vergence suggests an antiformal closure to the south-east. Such a closure could not, however, be the same as that deduced for the zone adjacent to Loch Affric because the plunge of the minor folds is re-

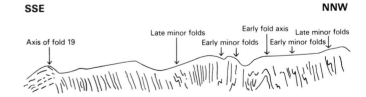

Figure 14 Sketch of cliff face [096 166] 360 m SSE of summit of Mullach Choire. Length of section about 600 m.

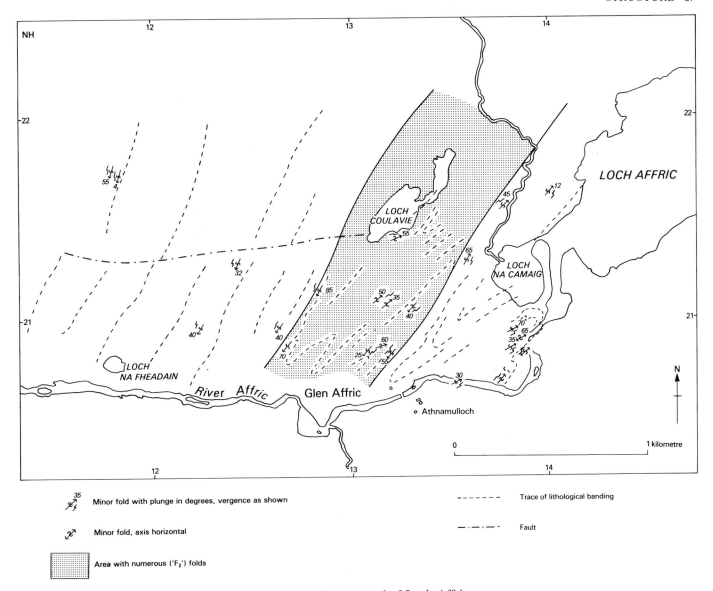

Figure 15 Sketch map of late ('F3') minor folds at the west end of Loch Affric.

versed. The consistency of plunge within the zones to the west and east of the central zone of 'F2' minor folds suggests that the 'F3' minor folds may have formed on an already tightly folded surface. Moreover, the vergence of 'F3' minor folds and the trends of lithological banding suggest that no single major structure dominates the area.

In the north-west part of sub-area D, in Glean a' Choilich, Choilich, west of Carn Eige, there is an area of complex structure with numerous minor folds resembling the 'F2' and 'F3' generations described above. The inter-relationship of these two fold phases on a large scale in this region is unclear, though both 'F2' and 'F3' major folds are present.

Minor 'F4' folds on the north side of Glen Affric (Figure 16e) clearly postdate the 'F3' folds. They trend north–south, are steeply dipping and have an associated crenulation cleavage in the pelitic bands.

Other late structures in sub-area D are faults developed along the steep limbs of folds. They are particularly well seen

adjacent to the stalker's path [146 208] where the vertical limbs trending about 20° are locally brecciated. In the same area a zone of brecciation with fragments of vein quartz and pegmatite as well as country rock can be traced for about 400 m in an easterly direction.

FOLDS OF INDETERMINATE AGE

The closure of fold 21 (Figures 6 and 16) terminates the outcrop of Psammite D to the north. Its near-vertical axial region (Figure 16g), developed in thick quartzites, is exposed on the north side of Gleann nam Fiadh. The fold cannot be traced to the south into the more-mixed micaceous psammite–semipelite lithologies. The possible northward and westward continuation of fold 21 may be represented by fold 21a (Figures 6 and 18). This east–west-trending structure refolds early minor folds and is accompanied by a tight crenulation cleavage, both features associated with D3 generation folds in sub-area F.

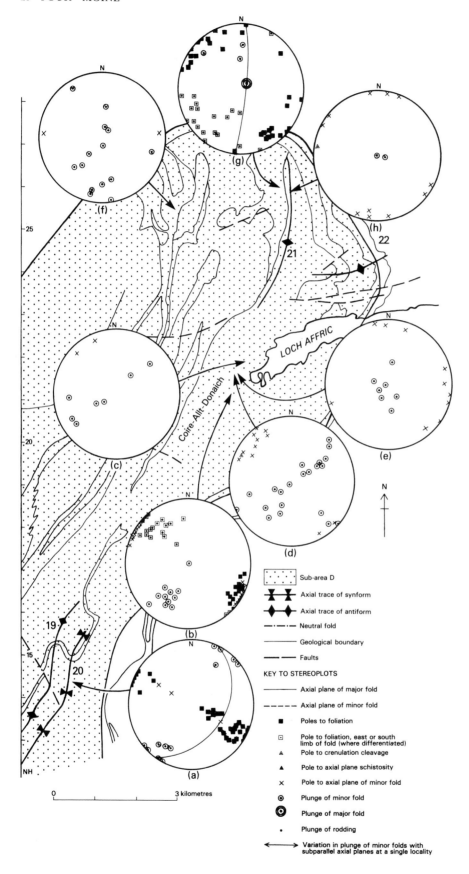

Figure 16 Stereographic projections of major and minor folds, sub-area D.

Table 7 Summary of data on major folds in sub-area D (see Figure 16)

Fold	Style	Axial-plane strike	dip	Plunge	Remarks
19 Antiform	Open	NNE	Moderate to ESE	Horizontal	Crenulation cleavage in pelite. Refolds early tight folds
20 Synform	Open	NNE	Moderate to ESE	Horizontal	As above. 19 and 20 form a fold pair with a subhorizontal common limb
21 Antiform	Tight to open	N	Steep to E	N to near vertical	A few associated open minor folds on N side of Gleann nam Fiadh. E limb refolded by open to tight almost vertically plunging minor folds trending E–W with crenulation cleavage. Local minor folds within bedding foliation
22 Antiform	Open	E–W	Almost vertical	Moderate	Associated open minor folds with coarse crenulation cleavage and local axial-lane schistosity. Refolds fold 23. Refolds tight minor folds with attenuated limbs

Sub-area E

This sub-area, extending eastwards from Loch Affric, includes a number of major folds (Figures 6 and 17). Its northern boundary is defined by the southern limit of tight west to east-trending late folds in sub-area F and by the general trend of foliation. The boundaries with sub-areas B, C and D have already been defined.

EARLY FOLDS

Tight to isoclinal minor folds, the axial planes of which are coincident with the trend of the regional lithological banding, occur in many areas. Their relationship to major folds is indicated in Table 8. The minor folds are characterised by attenuated limbs and, in micaceous rocks, by a dominant axial-plane schistosity. They refold bedding/foliation in psammite and a schistosity in bands of hornblende-schist. No major folds of this generation have been identified with certainty in sub-area E, but one possible structure is shown on Figure 6 and Figure 10d–f south of Loch Beinn a'Mheadhoin.

From the measurement of early fold profiles of thin psammite bands in semipelite at the bridge over the River Affric [283 283], just outside the Glen Affric district on Sheet 73W, the total strain can be assessed as at least 80 per cent displacement shortening; that is, the rock is reduced to 20 per cent of its original thickness in the plane normal to the fold axes. This degree of strain may be considered typical for mixed lithologies in lower Glen Affric. Similarly, post-D3 shortening has been estimated from buckle folds affecting pre-'D3' quartz and pegmatite veins to be about 50 per cent

in micaceous psammite near the western end of Loch Beinn a'Mheadhoin.

EARLY OR LATE FOLDS

Folds 23, 25 and 28 are tight to isoclinal major folds (Figure 17 and Table 8), but few minor structures appear to be associated with them. A schistosity in pelite or semipelite passes around the fold closures and there is no development of the marked axial-plane schistosity found in the early minor folds of the sub-area. Folds 25 and 27 are refolded by fold 26 (Figure 17). Fold 28 is probably an upward-facing anticlinal dome as described by Tobisch (1966), with the steeply south-plunging antiformal axis in the south turning over to become a synformal axis to the north. A north-trending crenulation cleavage is associated with minor open folds in mixed lithologies in its hinge area north of Loch Beinn a'Mheadhoin and may be associated with it. Tobisch regarded fold 28 as an F1 major fold refolded by F2, but Roberts et al. (1984) point out the similarity of folds in this area to single generation curvilinear 'sheath' folds described elsewhere in the Moine, features that suggest the fold is of the D3 generation.

LATE FOLDS

In the valley of the Allt Garbh, south of Loch Affric, generally eastward-plunging, open to very tight minor folds are characterised by an associated coarse crenulation cleavage or axial-plane schistosity in pelitic bands. They refold early 'F2' isoclines with highly attenuated limbs. The folds are accompanied by a rodding in quartz-feldspar lenticles and seem to postdate some larger bodies of deformed pegmatite [177

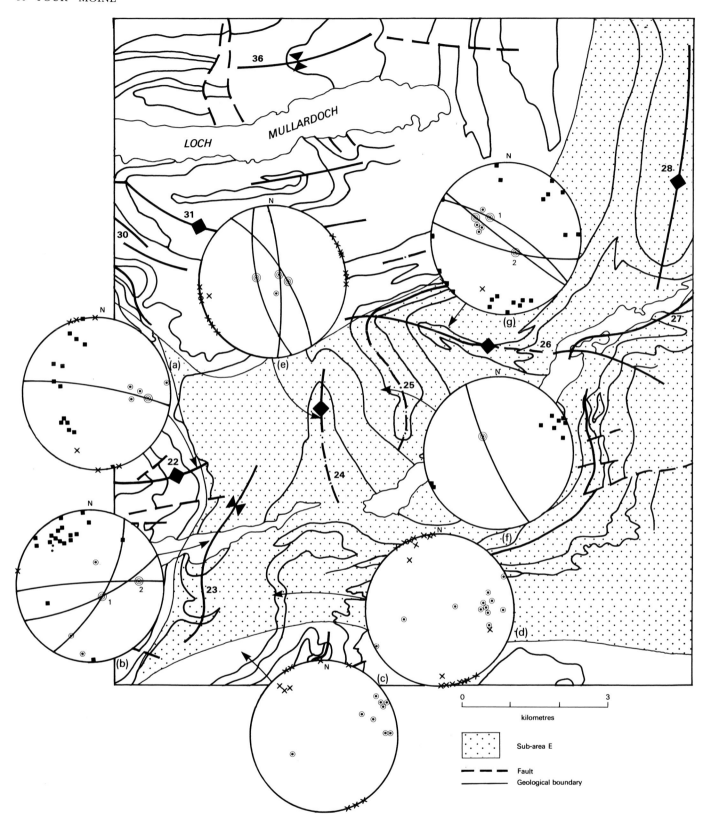

Figure 17 Stereographic projections of major and minor folds, sub-area E. For key to stereoplots see Figure 16, p.28.

203]. Plots of these minor folds (Figure 17 c and d), show that the vertical axial planes swing in azimuth from east-north-east to east in passing from the upper to the lower part of the Allt Garbh, possibly as a result of folding associated with fold 22 (see below). Minor folds of this style are also common in the complex folded mixed lithologies south of Loch Beinn a'Mheadhoin between Psammite B and G (Figure 3), where they follow the general trend of the foliation.

Fold 22 is an open upright structure associated with east-plunging open to tight minor folds which can be found over a wide area around Loch Affric and Loch Beinn a'Mheadhoin (Figure 17a). These are accompanied by a coarse crenulation cleavage and by rodding of quartz and pegmatite lenticles. The minor folds refold isoclines with attenuated limbs which can be referred to the early minor folds (D2) discussed above and postdate folds 23, 24 and 25. Hence, they may be post-

D3 as are folds 24 and 26 (Figure 17e and Table 8). These predate the minor structures associated with fold 22.

CORRELATION

The early folds are correlated on the basis of style and position within the overall structural sequence with the early folds of adjacent sub-areas and thus with the D2 generation folds of Roberts and Harris (1983). Fold 28 was referred to the early Cannich generation by Tobisch, (1966) and Tobisch et al. (1970), but this must be regarded as doubtful. Its general characteristics (Table 8) are those of D3 and its orientation is similar to that of D3 generation folds in the easternmost part of sub-area F. Fold 24 is similar in style to D3 folds in sub-area D but is slightly oblique to their typical north-north-east or north-east orientation. It is tentatively classed with the D3 generation, a correlation supported by its relationship to the early minor folds and to the structures as-

Table 8 Summary of data on major folds in sub-area E (see Figure 17)

Fold		Style	Axial-plane strike	dip	Plunge	Remarks
23	Sideways closing	Tight	NE	Steep to SE	Steep to SE	Few associated minor folds. Possible crenulation cleavage refolded by fold 22
24	Antiform	Open to isoclinal	NNW	Almost vertical	Steep to N or nearly vertical	Strong crenulation in pelite, associated tight to open minor folds. Folds early isoclines with axial-plane schistosity and attenuated limbs. Folded by minor folds associated with fold 22
25	Synform	Tight to isoclinal	NNW	Almost vertical	Steep to N	Poorly developed axial-plane schistosity in semipelite. Associated minor folds sparse or absent. Refolded by fold 26. Relationship to early minor folds not established
26	Antiform in W, synform in E	Open to tight	WNW	Almost vertical	Moderate to NW in W; vertical; moderate to SE in E	Crenulation cleavage and axial-plane schistosity in pelite. Associated minor folds and rodding common near axial trace. Refolds early minor folds with axial-plane schistosity
28	Synform in N, anticlinal dome	Tight	N	Steep to E	Steep to S	Crenulation cleavage locally developed. Schistosity in pelite folded by fold. No associated minor folds identified with certainty. Crossed by later local vertical E–W cleavage in pelite

sociated with fold 22. The crenulation cleavage and minor folds associated with fold 22 extend westwards into sub-area D where they post-date minor folds of D3 age (see above). Fold 26 is similar in style and orientation both to the east-ward-trending folds classed as D3 in sub-area F and to the post-D3 minor folds associated with fold 22 (see above). It refolds strata in the south-west extension of the closure zone of fold 28, which, as discussed above, seems to postdate D2. Fold 26 is therefore provisionally assigned to the same fold episode (post-D3) as fold 22.

Sub-area F

This sub-area extends to the north and north-west bound-aries of the map and is bounded to the south by a change from generally east–west-trending folds to more complex structures with more variable orientation (Figure 7).

EARLY FOLDS

Early tight to isoclinal minor folds ('F2') with axial-plane schistosity are common throughout the sub-area. They are preferentially developed in units where thin competent psammite horizons are interbanded with pelite or semipelite. In the east such structures fold cross-bedded psammite and the associated fabric is locally overgrown by muscovite porphyroblasts. On the south side of Loch Mullardoch the original bedding/schistosity in thicker pelite units is over-printed by the 'F2' schistosity and by subsequent pegmatite veining. Psammite bands are commonly boudinaged and complex interference structures ('F2'/'F3') are widespread (Plate 3). The 'F2' fold axial planes lie parallel to the regional foliation and their variation in attitude reflects swings in strike resulting from later folding (Figure 18a and b). The plunge of the axes shows a considerable spread, though generally plunging steeply south-west in the Beinn Fhionnlaidh area [116 283]. Farther west the 'F2' axial planes are parallel to the lithological banding which defines fold 34 and the axes plunge south-east (Figure 18f).

LATE FOLDS

The 'F3' minor folds postdate post-'F2' pegmatite veining which is locally well seen in pelite on the south shore of Loch

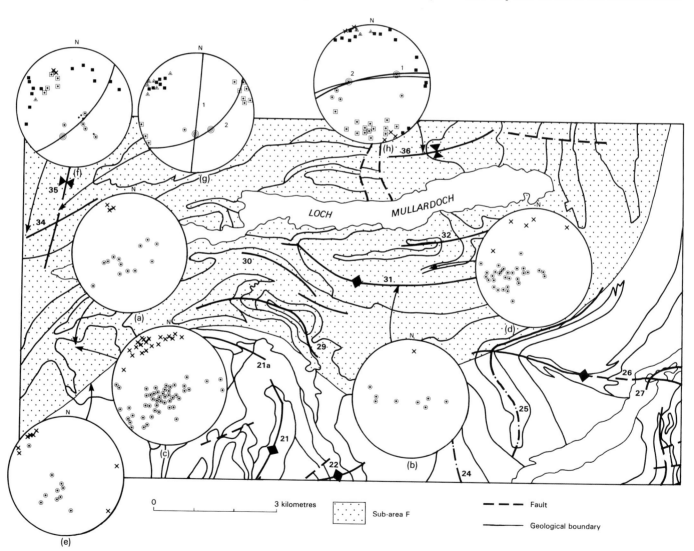

Figure 18 Stereographic projections of major and minor folds, sub-area F. For key to stereoplots see Figure 16, p.28.

Mullardoch (Plate 4). The 'F3' fold style varies from open to tight to rarely isoclinal and there is typically a related penetrative axial-plane schistosity or crenulation cleavage. In the area between Loch Mullardoch and Carn Eige [120 260] there is a series of recumbent, antiformal north-west-closing folds (for example folds 21a and 29) which fold the 'F2' fabric in their hinge zones. The intervening synformal structures are apparently absent, possibly as a result of sliding. Other major 'F3' structures are folds 30 to 34 and 36 (Figure 6). Although the plunge of 'F3' minor folds is generally steeply towards the south-west (Figure 18c and d) there is considerable local variation. The difference in attitude of the 'F3' folds shown in Figures 18c and d reflects the strike swings noted above. It is not known whether the 'F3' folds postdate or are superimposed on such variation.

An estimate of strain associated with 'F3' folds in thick psammite can be obtained from the difference in angles of cross-bedding across fold 31 (Figure 18). On the north limb measurements of the apparent dip of cross-bedding show a range of values from 11 to 18° (average 15°) and on the southern limb 10 to 31° (average 20). In the hinge, rota-tional effects give rise to cut-off angles of up to 90° between bedding and cross-bedding. Assuming flattening strain has occurred, the psammites appear to have been shortened to around 40–50 per cent of their original thickness.

Most of the area affected by fold 34 (Figure 6 and Table 9) lies west of the Glen Affric district. The fold varies from a north-east-trending, very tight antiform in the west to a steeply plunging almost sideways-closing structure where it enters the district (Figure 18f and g). Its eastern section is partly refolded by the open post-'F3' synform 35 (see below). Further east, fold 36 is a tight, basin-shaped structure, the axis of which varies through 100° in a distance of about 1.5 km (1 and 2 in Figure 18h). The widespread minor folds associated with folds 34 and 36 also locally show major varia-tions in plunge along their axial traces. A particularly good example occurs at a locality [175 315] where a saddle-shaped antiform can be traced for about 300 m in an east to west direction. Such plunge variations are common on both sides of Loch Mullardoch where they refold an earlier fabric in mixed lithologies.

Plate 3   Fold interference structures ('F2'/'F3') in a mixed sequence of pelite, psammite and semipelite with thin calcsilicate ribs. North side of Loch Mullardoch [217 315]. D3113.

Plate 4 Pegmatite veins in pelite and semipelite, all folded by 'F3' folds. South shore of Loch Mullardoch [178 302]. D3112.

Structures which postdate the 'F3' generation are developed only locally. In the north-west of the sub-area, fold 35 (Table 9, Figure 18g) is mainly developed in pelite. The associated lineation was noted on the north side of Loch Mullardoch as far east as the Allt Socrach [140 320]. On the south side of the loch post-D3 minor folds with an accompanying crenulation cleavage (Figure 18e) may be associated with the regional swings in orientation of 'F2' and 'F3' folds described below.

In the extreme east of the sub-area there is an abrupt swing in strike of the 'F3' minor folds from east–west to north–south. The initial part of the swing from east–west to north-east–south-west is shown by the trace of fold 36 (Figure 18), this possibly being the result of later folding.

CORRELATION

Folds 30 to 34 and 36 are believed to belong to the same generation ('F3') on the basis of style and continuity of minor structures. Fold 34 is a major fold which extends from the adjacent Kintail (72W) district where it can be correlated with folds such as the D3 Gleouraich Synform on similarity of style and trend (Figure 5). The curvilinear form of both major and minor folds is similar to that of D3 folds over a wide area further south (Roberts et al., 1984). The early minor 'F2' folds which are refolded by the D3 structures show similar features to the D2 folds described by Roberts and Harris (1983) and are thought to be the same generation.

Regarding correlation with areas to the north, fold 36 has been mapped as Orrin generation by Tobisch et al., (1970, Figure 2), which is one of several generations which they regarded as broadly post-D2 in age. The early minor folds equate in style with the Cannich generation (D2). No evidence has been seen for pre-Cannich generation folds (D1) which Tobisch et al., (1970) believed to be present at several localities around Loch Mullardoch, though pre-D2 syn-sedimentary slump folds have been recorded in psammite.

No firm correlation can be attempted for the major and minor folds which postdate D3. Fold 35 differs in trend and style from the Monar generation folds which occur a short distance to the north (Tobisch et al., 1970, Figure 2). The Monar folds trend north-east–south-west and have an associated axial plane schistosity. In the Glen Affric district these late folds show a more variable geometry and the coherent

Table 9 Summary of data on major folds in sub-area F (see Figure 18)

Fold		Style	Axial-plane strike	Dip	Plunge	Remarks
34	Antiform or sideways closing	Tight to open	NE	Steep to SE	Moderate SSW	Local crenulation cleavage, tight to open minor folds and associated rodding. Refolds early tight to isoclinal folds with axial plane schistosity. Refolded by fold 28
35	Synform	Tight to open	N	Approx. vertical	Steep to S	Crenulation cleavage. Accompanied by rodded pegmatite and quartz veins and mineral lineation. Folds earlier tight folds which fold pegmatite and quartz lenticles
36	Synform	Tight	E	Almost vertical	Steep to E and W	Crenulation cleavage and associated tight minor folds. Refolds early tight to isoclinal folds with axial plane schistosity. Locally refolded by vertical SW-plunging minor folds.

pattern of late folding seen around Loch Monar (Ramsay, 1958; Tobisch et al., 1970) does not continue southwards.

Conclusions

Figure 19 shows the axial traces of the well-defined major folds within the Glen Affric area and their proposed correlation. The area around the western end of Loch Loyne and Loch Cluanie is interpreted in the light of the work of Roberts and Harris (1983) and Roberts et al. (1984) which was based on the Loch Quoich area to south of Sheet 72E. The fold traces labelled post-D2 in the central part of Sheet 72E probably belong to the D3 episode but this requires confirmation. D2 and D3 structures are most easily recognised at the western end of Loch Cluanie and around Loch Mullardoch. Although the trends of these two sets of structures differ considerably some confidence is placed in this correlation. D2 minor structures are apparently present throughout the Sheet 72E area but it is difficult to trace D3 folds into the 'flat' belt. The full structural sequence of Tobisch et al. (1970), based mainly on work in the area north of the Glen Affric district, could not be substantiated during the current survey.

There are major swings in the regional strike of the banding and in the trends of the axial traces of D2 and D3 folds. These are most marked in the region between Glen Affric and Glen Cannich where the north-north-east-trending steep zone of Glenfinnan Division rocks turns to near east–west. The recognition of D3 structures here and their subsequent rotation suggests that the strike swing results either from later folding or reflects a pre-D3 geometry. However, minor structures related to these strike changes are only rarely seen and such swings may well be a consequence of deep-level

movements on faults in an underlying more rigid basement. Lewisian inliers become abundant immediately north of the Glen Affric district and Lewisian rocks may reasonably be assumed to underlie this area at a fairly shallow depth. In the Glen Affric area late east–west-trending folds (for example fold 22), which die out southwards, appear to reflect this swing in strike to the north. This is taken to imply that these late folds were formed by approximately north-south directed compression after the formation of the major strike inflexion. Further south it is unclear whether D3 folds change orientation to become more east- or north-east-trending, or whether further late east–west-trending folds exist and the D3 folds themselves die out into the 'flat' belt, as suggested by Roberts et al. (1984).

Post-D3 folds are only sporadically developed and no regional correlations can be made. The pattern of banding and fold trends in the 'flat' belt is controlled by late structures. Although the Cluanie Granodiorite could conceivably have caused some deflection of the structural traces (but see chapter six), it does not form a simple dome structure as the psammite to the north dips beneath it. The domal structure in the eastern part of the Glen Affric district (fold 12 on Figure 6) further influences the general trends of the banding and fold axes. This pattern is similar to that found in the 'flat' belt further south on Sheet 62E (Loch Lochy) where again local trends are largely controlled by domal structures.

In Glen Cannich and Strathglass in the adjoining Invermoriston district (Sheet 73W) the Glenfinnan Division rocks appear to grade stratigraphically upwards to the east into more psammitic rocks belonging to the Loch Eil Division. In most of the Glen Affric district, the distinction between the two divisions is more clearly defined. There is, however, some difficulty in defining the limits of Psammites F and G.

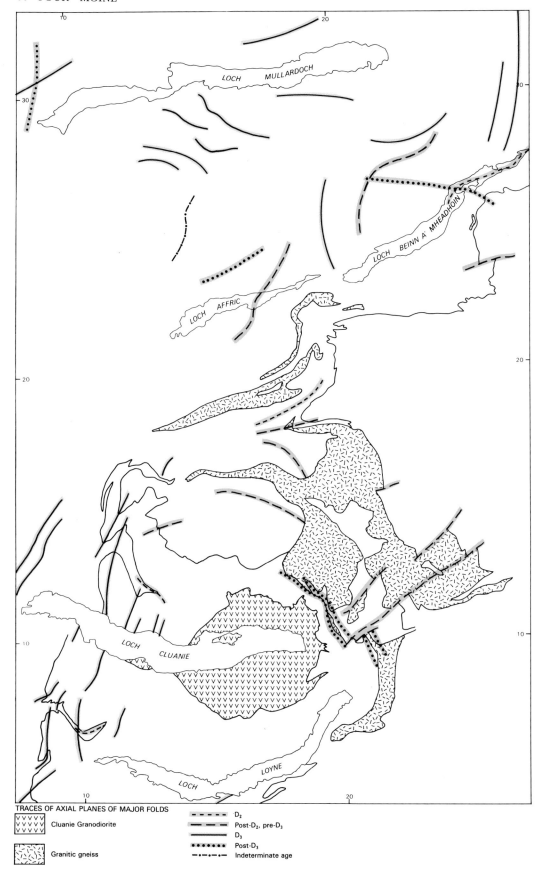

TRACES OF AXIAL PLANES OF MAJOR FOLDS

V V V V V V V V V V V V V V V Cluanie Granodiorite	⌐ ⌐ ⌐ ⌐	D₂
	▬ ▬ ▬	Post-D₂, pre-D₃
Granitic gneiss	▬▬▬	D₃
	••••••	Post-D₃
	▬·▬·▬·	Indeterminate age

Figure 19 Tentative interpretation of age relationships of major folds in the Glen Affric district.

Roberts and Harris (1983) equate these two psammites, as no intervening pelite units are seen around the western end of Loch Quoich. However, as shown on cross-sections in Figure 10, major D2 structures and at least one major slide zone make any stratigraphical correlation very tenuous. The latter dislocation is known to coincide in part with the western margin of the granitic gneiss sheets. The status of psammite F with its infolded tongues of Glenfinnan Division pelites is therefore uncertain as it cannot be confidently correlated with the Loch Eil Division (Psammite G). The locally contradictory younging evidence noted around the prominent pelite tongue north of Loch Cluanie (p.16) also implies that sliding is present further west, a feature confirmed by consideration of the structural geometry of the Glenfinnan Division rocks north of Glen Affric (Figure 10a–c). The area forming the north-west part of domal fold 12 (Figures 3 and 6) contains cross-bedding evidence implying that the sequence is inverted. If this inversion extends over a wide area, large-scale regional D2 folding and/or major sliding must be present, because to the east and north-east on Sheet 73W (Invermoriston) the Loch Eil Division rocks are largely right way up.

The distribution of pre-D2 metabasites (chapter five) across lithological boundaries of the Moine does suggest that relative movements on any slide zones have not been large enough to greatly disrupt the original pattern of lithologies. As there is no evidence for major thrusting during D1, it is concluded that the tectonic pattern formed by coherent major D2 and D3 folds has merely been modified by sliding, albeit locally quite strongly, during the D2 and D3 events. If the Moine is a stack of thrust sheets then the stratigraphical differences between structurally adjacent sheets can only be small. There is a concentration of metabasic rocks in the Loch Eil Division, particularly adjacent to the Glenfinnan/ Loch Eil Division boundary, but this may merely reflect an underlying lineament (Smith, 1979).

METAMORPHISM

The majority of the rocks within Sheet 72E are metamorphic and lie within the amphibolite facies. The age of metamorphism is uncertain but may relate to more than one major tectono-thermal event.

Alumino-silicate polymorphs within the Moine metasediments are rare and the metamorphic grade is generally indicated by the composition of plagioclase in calcsilicates (Fettes et al., 1984). The distribution of plagioclase values within the sheet shows two distinct zones (Figure 20), an easterly zone with values of $An_{30} - An_{60}$ and a westerly zone with values of $> An_{60}$. Intermediate values are confined to a narrow belt between the two zones, reflecting the rapid change in plagioclase composition with increasing grade. The zonal pattern reflects the position of the Glen Affric district on the eastern flank of the spine of high metamorphic grade which runs approximately north-east–south-west across Western Inverness-shire (Figure 20). Fettes et al., (1984) assigns the easterly zone to the lower amphibolite facies, which in pelites is characterised by kyanite + staurolite assemblages. The westerly zone is assigned to the middle amphibolite facies characterised by sillimanite + muscovite assemblages in pelites.

Pelitic and semipelitic rocks

The general lithology of the pelitic and semipelitic rocks has been described elsewhere (p.11). Across the sheet they present the following relatively uniform assemblage:

quartz-muscovite-biotite-plagioclase-(garnet)-(potash feldspar)-(kyanite)

Kyanite porphyroblasts have been described by Tobisch (1963) from Glen Cannich. Relict kyanite porphyroblasts surrounded by shimmer aggregate have been noted at [1928 2956]. Muscovite is commonly the dominant mica and may occur either as plates defining the general foliation of the rock or as masses of shimmer aggregate possibly after alumino-silicate porphyroblasts. In addition muscovite is commonly found as large discrete porphyroblastic plates (see p.11). Biotite occurs as plates intergrown with the muscovite, both micas defining the main foliation of the rock. Garnets occur either as small subhedral grains clear of inclusions, or as larger corroded crystals with scattered small inclusions. Rarely garnets are found exhibiting a clear zonation marked by a variation in the density or style of inclusions.

The garnet, muscovite and kyanite porphyroblasts, as well as the ovoids of shimmer aggregate, are all wrapped by the main foliation. There is very little evidence of porphyroblastic crystallisation after the development of this foliation, although the possibility of some garnet growth and development of shimmer aggregate cannot be ruled out.

Muscovite and biotite compositions are relatively uniform across the sheet; average analyses are given in Table 10. Biotite M/MF (100 Mg/Mg + Fe) values range from 17 to 29. Even within individual thin sections the biotites adjacent to garnets have identical compositions to those in the general fabric.

The garnets are almandine-rich with compositional zoning generally being confined to a narrow margin, giving typically flattish chemical profiles. Representative compositions are given in Table 11.

Garnets with textural zoning also show little sign of compositional variation; for example, in S 64361* [1234 2622]

Inner Zone	Outer Zone	Rim
$Py_{14}Al_{77}Sp_4Gr_5$	$Py_{13}Al_{78}Sp_3Gr_6$	$Py_{11}Al_{80}Sp_3Gr_6$

This is in contrast to those garnets examined from the Moine pelites and semipelites lying between the Glen Affric district and the Great Glen where pronounced compositional profiles are characteristic.

Calcsilicate rocks

The lithology and occurrence of calcsilicates has been described elsewhere (p.8). They may show a considerable mineralogical variation and are commonly zoned with transi-

---

* Rock slice number in British Geological Survey collection

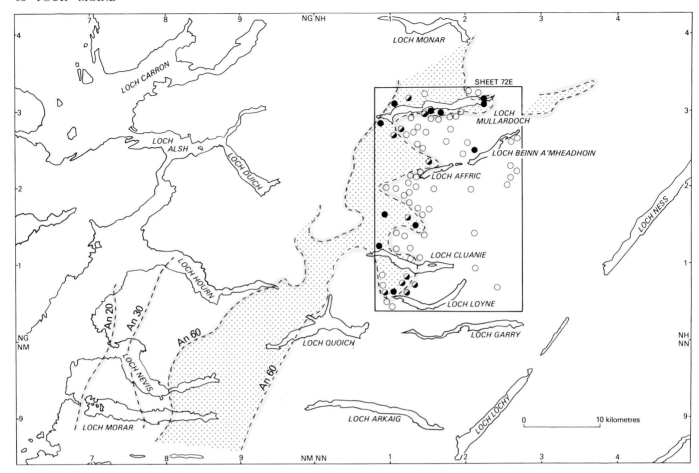

Figure 20 Regional setting of Sheet 72E in terms of the anorthite content of plagioclase in calc-silicate rocks. The An_{20} An_{30} and An_{60} isograds are shown, with the high grade zone stippled. Within the area of Sheet 72E individual measurements of anorthite percentage are shown thus:
open circles, values of <50%
half solid circles, values of 50–70%
solid circles, value of >70%.

Table 10 Analyses of micas from Moine rocks of Sheet 72E

	1	2	3
SiO_2	35.11	36.27	45.45
TiO_2	1.78	2.01	1.10
Al_2O_3	18.83	17.79	33.29
FeO	21.33	19.51	1.72
MnO	0.10	0.42	—
MgO	8.36	9.64	0.72
CaO	0.01	0.06	0.01
Na_2O	0.12	0.00	0.72
K_2O	8.40	9.12	10.13
M/MF	23.31	27.70	

1) Average biotite analysis from ten pelites and semipelites
2) Average biotite analysis from two calcsilicates
3) Average muscovite analysis from five pelites and semipelites

tional facies against the host-psammites and semipelites. The assemblages of Moine calcsilicates have been extensively studied, for example by Tanner (1976) and Winchester (1972). Within the area of sheet 72E the main assemblages are:

quartz-biotite-plagioclase-(hornblende)-(garnet)-epidote/zoisite;
quartz-(biotite)-plagioclase-hornblende-(clinopyroxene)-(garnet)-(epidote/zoisite);
quartz-zoisite.

Pyroxene is confined to a few localities in the west of the sheet. The plagioclase values vary across the sheet as indicated on Figure 20. Zoned plagioclases have been recorded within the eastern area. These show calcic-rich cores; for example, in S 65081 [2330 3123] a variation of $Or_0Ab_5An_{85}$ to $Or_1Ab_{54}An_{45}$ has been measured.

Regionally the relationship of these assemblages to the main foliation and deformational fabrics is uncertain.

Table 11 Compositional variation in garnets from
Moine rocks of Sheet 72E

Core	Rim
1) Py_{13} Al_{78} Sp_3 Gr_6	Py_8 Al_{80} Sp_5 Gr_7
2) Py_{10} Al_{76} Sp_6 Gr_4 An_2	Py_4 Al_{55} Sp_{14} Gr_{25} An_2
3) Py_{12} Al_{82} Sp_2 Gr_3 $An1$	Py_{11} Al_{84} Sp_2 An_1

1) Average garnet composition from 11 pelites and semipelites

2) Example of strong MnO and CaO enrichment of the garnet rim in a
quartz-biotite (M/FM = 17)-muscovite-garnet-plagioclase assemblage,
S 63172 [2341 2841]

3) Example of lack of compositional change between the garnet rim and
core in a quartz-biotite (M/FM = 25)-muscovite-garnet-oligoclase-kyanite
assemblage, S 65254 [1928 2956]

Metamorphic history

Textural observations from thin sections of pelites indicate
that the main phase of porphyroblast growth predated the
development of the main foliation (D2). This late, penetrative, foliation is defined by the parallel alignment of white
mica and biotite. The composition of the latter is uniform
across the sheet area (and further east), even where in contact with garnet. This contrasts with the compositional profiles across individual garnets, which are typically flat within
the Glen Affric district, but variable farther east. Such
evidence suggests a prolonged thermal event or perhaps a
later reheating resulting in equilibration of individual garnet
compositions.

The events reflected in the pelites cannot be easily related
to those which resulted in the zonation indicated by the calcsilicates. The appearance of normally zoned plagioclase in
the eastern part of the district may indicate a zone of retro-
gression relative to the west.

FIVE

Pre-Caledonian igneous rocks

GRANITIC GNEISS

The granitic gneiss is here classified as an early, pre-orogenic granite, in agreement with Harris (1983), and Roberts and Harris (1983), although it has also been suggested that it is a migmatite (Dalziel, 1966) or a slice of basement tectonically introduced into the Moine rocks (Harris in discussion of Winchester, 1974). It crops out near or at the contact between the Glenfinnan Division and the Loch Eil Division (Figures 2 and 21). It is typically a coarse-grained, pinkish, white-weathering rock with a gneissosity imparted by alignment of biotite and by segregation of the coarser-grained feldspar and quartz. At some localities the rock is massive and granite-like, for instance, that forming much of the sheet which crops out in Coire Dho [19 13], but elsewhere it is well foliated. Finer-grained varieties resemble coarse, massive psammite. Felsic lit-par-lit segregation of quartz and microcline occurs locally along the margins of the granitic gneiss sheets.

Thin sections show the gneiss to be composed predominantly of varying proportions of K-feldspar and quartz. Plagioclase (oligoclase to low andesine) forms 10 to 40 per cent of the rock and micas (biotite and subsidiary muscovite) about 5 per cent. Garnet is a widespread accessory mineral, together with a little epidote, sphene, apatite and iron oxide. In the coarser, well-foliated rock, the larger feldspars tend to be porphyroclastic, set in a finer-grained matrix. Hornblende is present in the gneiss of at least two localities [177 196] and [200 173] and may support its igneous parentage (Rock, 1983).

Granitic gneiss can be traced northwards from the Loch Lochy (62E) Sheet as discontinuous bands. On the southeast side of Loch Loyne the bands occur entirely within psammite, but northwards from the east end of Loch Cluanie the western margin of a folded sheet (Figure 3) cross-cuts the lithological banding of the country rock on a large scale (see below). In Glen Moriston a tongue of gneiss extends at least as far east as Tomchrasky [255 123] and may be continued in the area of the adjoining Invermoriston (73W) map. Between the River Doe and Glen Affric the sheets of gneiss emplaced in pelite are folded by the major late post-D2–pre-D3 folds. South and east of Loch Beinn a'Mheadhoin the granitic gneiss occurs as lenses, the most northerly of which is immediately west of Loch Carn na Glas-leitire [255 252].

The contacts of the granitic gneiss are commonly sharp, but may be gradational. Examples of the former can be seen where the gneiss is in contact with pelite above the Ceannacroc power station tailrace tunnel [227 113] and with psammite in the River Loyne [217 098]. There are examples of gradational contacts higher up the course of the River Loyne [213 087] and in Gleann Fada. At the latter locality [174 157], psammite and semipelite within 100 m of the con-

tact contain numerous concordant pegmatite veins. Similar veins which occur in the gneiss within a short distance of the contact are either concordant or slightly discordant to the foliation (cf. Dalziel 1966, Plate 2, Figure 2).

Transgressive relationships of the granitic gneiss to the enclosing rock occur on both large and small scales, though for the most part the observable junctions are concordant. For instance, as mentioned above, the gneiss cross-cuts the lithological banding of the country rock on a large scale north of Loch Cluanie, although in individual exposures the relationship appears concordant. At a locality [134 028] just outside the southern border of the map, the granitic gneiss transgresses horizons of psammite on a small scale and also sends off a cross-cutting sheet (Figure 22). The former type of transgression can be explained either as the result of differential movement between the gneiss and its envelope of country rock resulting in sliding (see p.24) or, less likely, as an original feature of its emplacement. The small-scale cross-cutting relationship suggests local mobilisation and intrusion of the gneiss, possibly at a later date.

Apparently concordant intercalations of metasediments are interbanded on a small scale with granitic gneiss within and at the margin of some of the major sheets; for instance, those in the area west of Ceannacroc Lodge. Here, at the south margin of the sheet adjacent to the Allt Ruigh Fhearchair [202 106], the gneiss is interbanded with pelite and psammite. An enclave [217 115] within another sheet consists of interbanded pelite, psammite and granitic gneiss.

Small-scale fold structures are well displayed in the gneiss at some localities but are obscure in the more massive varieties. In the stream section of the lower River Doe the following structural history can be made out:
1. Foliation of gneiss
2. Emplacement of pegmatite veinlets which locally cross-cut 1
3. Folding of 1 by tight to isoclinal curvilinear folds (D2) accompanied by locally extreme attenuation of the fold limbs. Imposition of an axial plane schistosity, rodding and mullions.
4. Open folding

On a larger scale the granitic gneiss in the Glen Moriston area is apparently disposed as an isoclinally folded sheet (Figure 10g–h), the isocline being interpreted as a D2 fold, perhaps similar to those which apparently deform the gneiss in Ardgour (Dalziel, 1966).

As described below, the granitic gneiss in the River Doe area was intruded by metabasites before D2 and there is a suggestion of a long geological history prior to this event. No age dates have been obtained for the granitic gneiss within the Glen Affric district, but at Glenfinnan on the Loch Quoich (62W) Sheet, an $^{87}Rb/^{86}Sr$ isochron for one of the more southerly sheets of the gneiss yielded an age of 1056 ± 46 Ma (Brook et al., 1976).

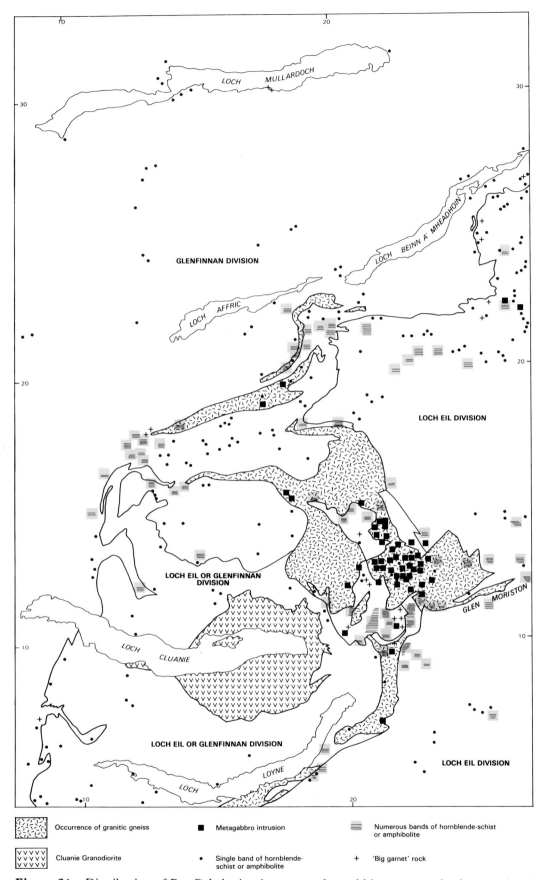

Figure 21   Distribution of Pre-Caledonian igneous rocks and big garnet rock of uncertain origin.

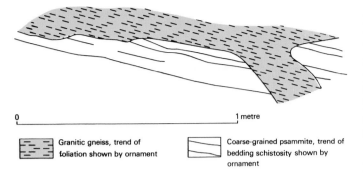

| Granitic gneiss, trend of foliation shown by ornament | Coarse-grained psammite, trend of bedding schistosity shown by ornament |

Figure 22 Offshoot of granitic gneiss cutting across psammite. Vertical, west-facing section at [134 028].

METABASITE

General

Throughout much of the area under discussion the Moine metasedimentary rocks are accompanied by, and are locally cut by, bodies of metabasite (the amphibolite suite of Smith, 1979) which vary in size from stripes several millimetres in thickness to pods and sheets many metres thick. They are interpreted as metamorphosed minor intrusions. A few occurrences, however, may be allied to the big garnet rocks described in chapter four. They are to be found for the most part within a few kilometres of the contact of the rocks of the Glenfinnan Division with the Loch Eil Psammite (Figure 21), and are particularly common towards the eastern limit of the granitic gneiss outcrops.

The metabasites comprise metagabbro, which is characterised by a well-preserved macroscopic gabbroic or doleritic igneous texture, massive amphibolite and hornblende-schist (Figure 21). In the amphibolites and hornblende-schists the fabric is thoroughly metamorphic. Most of the metabasites take the form of concordant sheets or irregular masses which range from about 20 m thick to a centimetre or less. A few are cross-cutting, but it is not always possible to determine whether this is an original intrusive feature or the result of shearing during deformation. As discussed below, the metabasites were intruded prior to D2 and locally comprise more than one generation. Preliminary Rb-Sr dating by the Isotope Geology Unit of BGS of a large metagabbro body in the bed of the River Doe [219 126] and of smaller bodies around Cannich in the adjacent Invermoriston district is consistent with a Precambrian age of intrusion (M Brook, personal communication 1983; Rock et al., 1985).

The hornblende-schists, amphibolites and metagabbros are mineralogically simple, being composed chiefly of greenish brown hornblende and plagioclase with usually minor amounts of quartz, sphene, epidote, biotite, garnet and opaque oxides. Garnet, sometimes occurring as megacrysts, is prominent in some rocks and locally overprints an earlier fine-grained fabric. The lineation in the more schistose rocks is defined by hornblende and the plagioclase commonly shows a strong dimensional orientation. The metagabbros exposed in the valley of the River Doe (Peacock, 1977) include feldsparphyric varieties, for example at [224 125], and in a number of cases retain the apparently unchanged igneous plagioclase with an anorthite content of 50 per cent. Elsewhere in metabasites the anorthite content of the plagioclase is usually lower. In some rocks the hornblende is replaced by biotite and the plagioclase is partly replaced by epidote, biotite, hornblende and garnet. The primary igneous feldspars in the metagabbros of the River Doe are locally clouded and occasionally show discontinuous reaction rims of garnet against the mafic minerals (Peacock, 1977, p.3).

Distribution

The intrusive metabasites occur chiefly in the south-eastern part of the Sheet 72E (Figure 21). Though the distributions of garnet-rich and garnet-poor varieties differ, there is insufficient evidence to show that they belong to different suites.

The intrusive metabasites are most numerous in and adjacent to the outcrop of the granitic gneiss and within a few kilometres of the boundary of the Loch Eil Psammite. They are also abundant along the west margin of Psammite F. Within these broad limits of distribution there may be areas of greater and lesser concentration but these are not well defined because of variations in bedrock exposure from place to place. As the broad belts of numerous metabasites straddle major lithological boundaries it can be inferred that such boundaries have not been modified extensively on a regional scale by tectonism postdating the period of intrusion prior to D2 (see also p.37, chapter four).

Field relations

In the south-east part of the Glen Affric Sheet it can be demonstrated that the metagabbros and many of the hornblende-schists belong to a single suite of minor intrusions that has been deformed to varying degrees (Peacock, 1977). The metagabbros, of which about 50 have been mapped in the valley of the River Doe (Figure 21), form for the most part concordant pods several metres across within the outcrop of the granitic gneiss, but one exceptionally large body occupies the bed of the River Doe at [2190 1264] for a distance of about 70 m (Figure 23). The body is a multiple intrusion with early partly banded gabbroic metabasite (B) containing xenoliths of granitic gneiss. This is cut by a pegmatitic metagabbro (C) which is in turn cut by a multiple sheet-like intrusion of doleritic metabasite (D). The variable-textured metabasite (A) predates (D) but its age with reference to B and C is not clear. Analyses of these rocks are given in Table 12. Many of the other metagabbro bodies, for example that seen by the roadside at [2108 1262], include later doleritic veins.

Transitions between metagabbro and hornblende-schist, and the relationship of these rocks to the fold history in the granitic gneiss, are well seen in the bed of the River Doe and tributary streams. On the east side of the fault in the lower course of the Allt Bhruisgidh [2175 1267] a body of metagabbro shows a gradation from a rock with a well-preserved relict igneous texture to hornblende-schist in which the gabbroic texture is progressively streaked out and destroyed. Upstream, in the bed of the River Doe [2166 1266] bands of hornblende-schist identical to the more deformed parts of

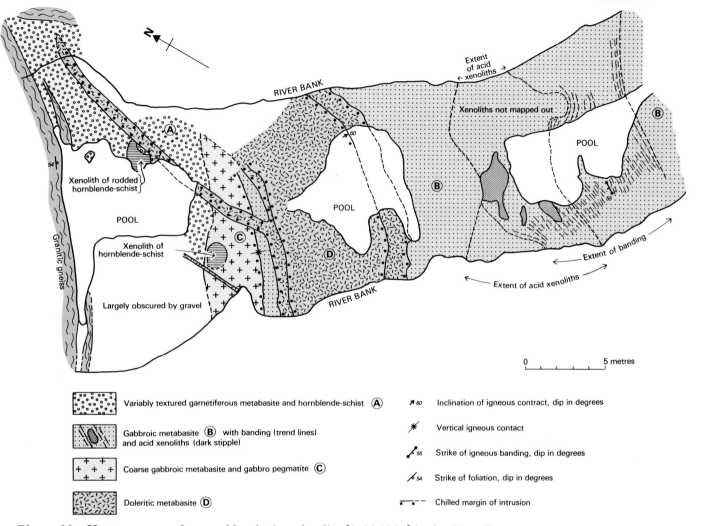

Figure 23   Upstream part of metagabbro body at locality [2190 1264] in the River Doe.

this metagabbro are interleaved and folded with granitic gneiss. The folds (D2), which vary from isoclinal to fairly tight, are accompanied by a strong mineral lineation in both rock types (Figure 12 c and d). In many folds, a strong axial-plane schistosity is developed in both hornblende-schist and granitic gneiss. Compositional banding in hornblende-schist is folded by one such fold, which also folds a schistosity in the schist. At another locality [2097 1280] a pod of metagabbro occupies the core of a flat-lying isoclinal fold (D2) in the granitic gneiss (Figure 24) and can be traced as a thick sheet downstream for over 50 m. The margin of the pod is partly schistose, especially at the fold hinge, and this schistosity can be traced round the fold closure. Within the metagabbro, veinlets of acid rock compositionally identical to the granitic gneiss envelope are rodded parallel to the D2 fold axis, and a strong lineation due to extension of the subophitic texture is locally developed in the metagabbro, this also being parallel to the fold axis.

Further north, metagabbros are present in the Loch Eil Psammite close to its margin with rocks of the Glenfinnan Division. Two of these occur within Sheet 72E (Figure 21) and there are others in the valley of the River Affric east of Loch Affric. Though these intrusions retain a megascopic ig-

neous texture they are thoroughly recrystallised. They locally cut the banding in the psammite (Figure 25) in an area where sedimentary structures are preserved, suggesting that the cross-cutting relationships are due to igneous intrusion rather than to shearing.

The field relationships of the hornblende-schists and amphibolites are similar in some respects to those of the metagabbros. The larger masses generally occur as pods with a marginal schistosity conformable to the country rock foliation, though cross-cutting relationships can be seen locally (Figure 26).

The probability that there are two suites of hornblende-schists, one being earlier than the metagabbros and the other contemporaneous with them, is suggested by the evidence at two localities in the River Doe. At the north end of the large metagabbro intrusion shown in Figure 23, xenoliths of horn-blende-schist can be seen in both the pegmatitic facies (C) and the garnetiferous, partly schistose metabasite (A). In view of the paucity of xenoliths in the metagabbros as a whole, it is probable that the fragments of hornblende-schist are of local derivation and are thus highly unlikely to be far-travelled clasts of, say, Lewisian basement. Relationships further upstream, at [2108 1270] are shown in Figure 27.

Table 12   Analyses of metabasites and big garnet rock of uncertain origin (analyst B Walker)
Nos. 1 to 8. Large metagrabbro body exposed in River Doe [219 126]

Sample No.	1	2	3	4	5	6	7	8	9	10	11	12	13	14	15	16	17
SiO_2	47.7	48.4	48.7	48.0	48.2	54.2	53.4	50.2	48.3	48.7	48.4	47.6	48.5	48.3	48.6	49.7	53.6
TiO_2	4.43	4.15	4.04	1.71	1.76	3.24	4.32	3.39	1.30	4.29	3.63	2.03	1.63	1.96	1.36	1.42	2.40
Al_2O_3	12.87	12.98	13.21	16.11	15.61	12.58	12.87	12.83	15.75	12.65	12.66	14.96	17.61	15.01	15.97	15.07	11.27
Fe_2O_3	2.72	2.37	2.47	1.60	1.21	3.12	2.15	3.07	1.49	3.84	2.85	2.24	1.45	1.93	1.71	1.54	6.25
FeO	12.71	12.20	12.24	9.01	9.63	9.86	10.48	11.71	8.90	12.05	12.70	10.30	9.07	9.63	8.53	8.57	12.96
MnO	0.23	0.26	0.25	0.20	0.22	0.23	0.19	0.23	0.21	0.30	0.29	0.24	0.24	0.22	0.19	0.17	0.30
MgO	5.36	5.08	5.53	8.28	8.06	4.20	3.71	4.80	8.24	4.70	5.23	7.14	7.03	7.60	8.39	8.29	5.73
CaO	9.42	8.45	9.21	11.13	11.16	7.70	7.64	8.03	11.31	8.99	9.27	10.78	8.89	10.51	11.41	11.72	1.83
Na_2O	2.54	2.39	2.27	2.83	2.71	2.86	2.24	2.61	2.42	1.94	2.16	2.45	2.60	2.54	2.20	2.19	0.06
K_2O	0.88	1.13	0.91	0.43	0.54	0.65	1.35	1.10	0.77	0.99	0.90	0.68	1.72	1.00	0.73	0.46	3.06
P_2O_5	0.44	0.46	0.46	0.25	0.26	0.46	0.29	0.53	0.22	0.51	0.57	0.27	0.25	0.27	0.24	0.24	1.13
H_2O	1.18	1.60	1.16	1.21	0.89	0.73	0.82	1.12	1.26	1.09	1.47	1.16	0.95	1.48	1.35	1.27	1.20
Total	100.48	100.37	100.45	100.76	100.25	99.83	99.46	99.62	100.17	100.05	100.13	99.85	99.94	100.45	100.68	100.64	99.79
Ce	46	60	41	9	12	50	62	64	<1	46	53	30	9	3	19	3	70
Nb	9	8	8	<1	3	10	12	9	2	13	13	3	3	5	4	2	16
Rb	19	37	27	17	22	10	41	40	39	34	24	14	97	42	27	16	119
Sr	142	173	142	174	192	153	134	170	181	133	149	152	369	215	151	161	50
Y	67	74	70	30	31	75	62	100	27	83	88	36	31	34	29	28	102
Zr	269	299	276	97	105	277	205	413	72	320	374	119	95	108	79	82	394
Cr	—	119	—	251	180	108	—	127	139	—	41	89	—	—	201	366	72
Ni	43	35	50	100	66	34	22	38	60	34	31	49	66	59	71	62	15
Ba	143	309	183	<50	84	150	242	276	78	189	108	<50	224	87	53	59	523
Pb	7	9	4	2	3	10	9	15	1	7	5	5	<1	2	<1	3	7
Zn	117	299	124	66	70	103	116	139	59	130	141	81	82	80	66	70	170
Cu	31	31	33	43	62	23	8	26	14	45	40	51	46	35	22	26	18
Ce/Y	1.74	2.06	1.67	0.76	0.98	1.07	2.54	1.63	<0.09	1.41	1.53	2.11	0.74	0.22	1.69	0.27	1.74
Rb/Sr	0.13	0.21	0.19	0.09	0.11	0.06	0.30	0.23	0.21	0.25	0.16	0.09	0.26	0.19	0.18	0.10	2.4

Nos. 1 to 8   Large metagabbro body exposed in River Doe [219 126]
Reference No.
1   S54716   Metagabbro without layering (B on Figure 23)
2   GD1e   Metagrabbro without layering (B on Figure 23)
3   S 54715   Layered metagrabbro (B on Figure 23)
4   S 54012   Metadolerite (D on Figure 23)
5   GD1d   Metadolerite (D on Figure 23)
6   GD1a   Variably textured garnetiferous metabasite and hornblende-schist (A on Figure 23)
7   S 54713   Variably textured garnetiferous metabasite and hornblende-schist (A on Figure 23)
8   GD1g   Xenolith of hornblende-schist (C on Figure 23)

Nos. 9 to 15   Various metagrabbros and hornblende-schists, River Doe area
Reference No.
9   GD2   Metagabbro. Mouth of Allt Bhuruisgidh [217 127]

10   S 54008   Hornblende-schist with large sieve garnets [218 126] (Figure 28)
11   GD3a   Hornblende-schist with rodded aggregates [211 127] (Figure 27)
12   GD4a   Metagabbro [209 128] (Figure 25)
13   S 54946   Feldsparphyric metagabbro [224 125]
14   S 54945   Sparsely porphyritic metagrabbro [226 128]

Nos. 15 and 16   South-west of Hilton Lodge [285 246], Sheet (73W) Invermoriston
Reference No.
15   S 64996   Metagabbro [281 240]
16   S 64995   Metagabbro [281 240]
Reference No.
17   S 59238   Big garnet rock of uncertain origin

Metagabbro with a fine-grained chilled margin apparently cross-cuts both the granitic gneiss and a concordant hornblende-schist which is characterised by rodded aggregates of quartz, plagioclase and garnet up to 15 cm long. Interleaved non-garnetiferous hornblende-schist, which is indistinguishable from that associated with the metagabbros, is also apparently cut by the metagabbro, though the actual contact is obscured. However, at another locality in the same river section, at [2183 1262] (Figure 28) garnetiferous hornblende-schist similar to the earlier hornblende-schist illustrated on Figure 27 apparently cuts non-garnetiferous hornblende-

schist identical to that associated with the metagabbros of the area. These exposures, of hornblende-schist in metagabbro and one schist cross-cutting another, point to the existence of two generations of metabasite.

The geological history of the metabasites in the River Doe area is interpreted as follows (slightly modified from Peacock (1977)):

a. Foliation in granitic gneiss;probable intrusion and metamorphism of earlier generation of basic rocks and their metamorphism to hornblende-schists.

b. Intrusion of metagabbros and associated hornblende-

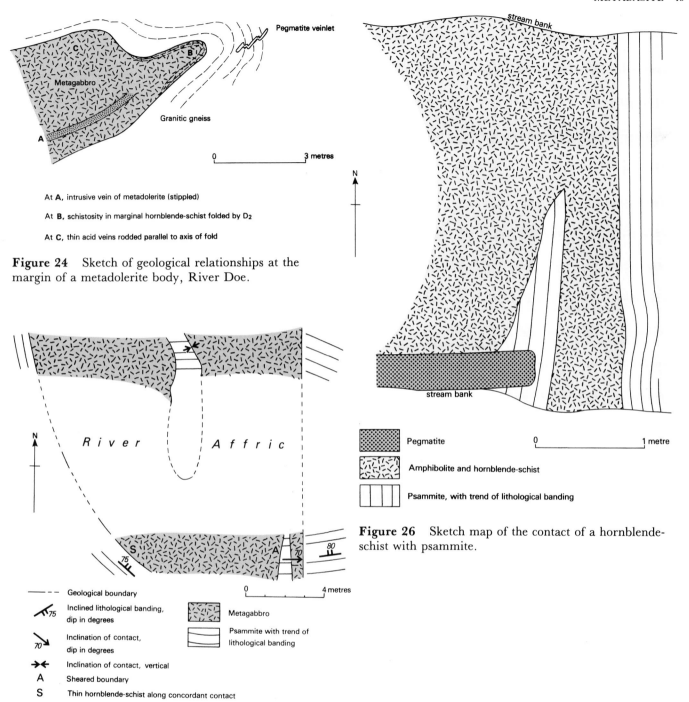

Figure 24 Sketch of geological relationships at the margin of a metadolerite body, River Doe.

At **A**, intrusive vein of metadolerite (stippled)

At **B**, schistosity in marginal hornblende-schist folded by D2

At **C**, thin acid veins rodded parallel to axis of fold

Pegmatite veinlet

Metagabbro

Granitic gneiss

0   3 metres

N

River   A f f r i c

Geological boundary

Inclined lithological banding, dip in degrees

Inclination of contact, dip in degrees

Inclination of contact, vertical

A   Sheared boundary

S   Thin hornblende-schist along concordant contact

Metagabbro

Psammite with trend of lithological banding

0   4 metres

Figure 25 Sketch map of metagabbro intrusion, River Affric.

stream bank

stream bank

Pegmatite

Amphibolite and hornblende-schist

Psammite, with trend of lithological banding

0   1 metre

Figure 26 Sketch map of the contact of a hornblende-schist with psammite.

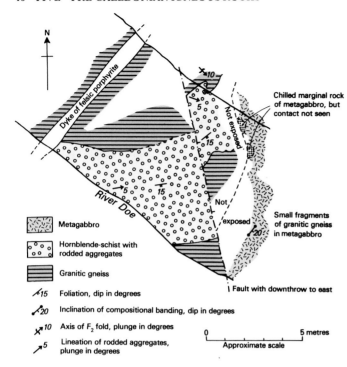

Figure 27 Sketch map of geological relationships of metabasites with granite gneiss, River Doe.

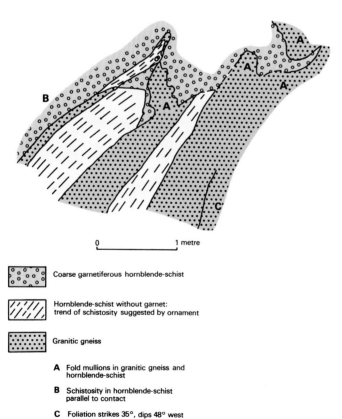

Figure 28 Sketch of apparently cross-cutting hornblende-schist, River Doe.

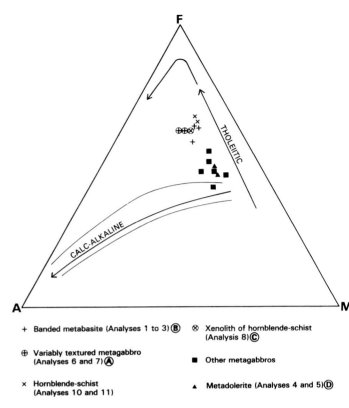

Figure 29 Metabasites, AFM plot. A, B and C are from units indicated on Figure 23.

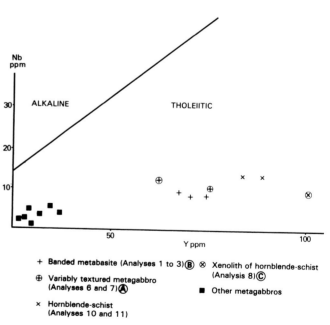

Figure 30 Metabasites, plot of Nb against Y. A, B and C are from units indicated on Figure 23.

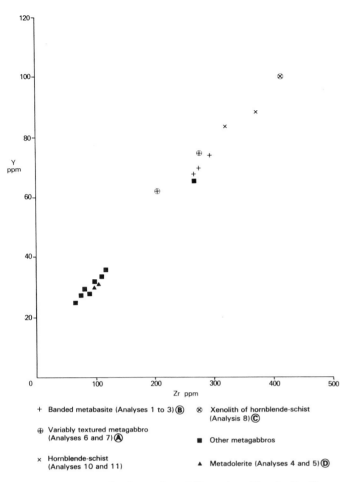

+ Banded metabasite (Analyses 1 to 3) Ⓑ
⊕ Variably textured metagabbro (Analyses 6 and 7) Ⓐ
× Hornblende-schist (Analyses 10 and 11)
⊗ Xenolith of hornblende-schist (Analysis 8) Ⓒ
■ Other metagabbros
▲ Metadolerite (Analyses 4 and 5) Ⓓ

Figure 31 Metabasites, plot of Y against Zr. A, B, C and D are from units indicated on Figure 23.

schists; imposition of 'additional' schistosity.

c. Tight to isoclinal folding with development of axial plane schistosity accompanied by strong lineation, rodding and mullion structures (D2).

d. Open folding.

This suggests a long geological history within the granitic gneiss outcrop prior to D2. Peacock (1977) suggested that the 'additional' schistosity of item b could have been imposed almost *pari passu* with the intrusion of the metagabbros and was analogous to the schistosity/foliation found in members of the late Caledonian microdiorites (see p.56). Alternatively, if it were formed during a discrete tectonic event (D1 of Holdsworth and Roberts, 1984) the history of the granitic gneiss would include an earlier tectonic episode for which no evidence has been found elsewhere within this part of the Moine outcrop.

Chemistry

Analyses of a number of metabasites, chiefly from the River Doe area, are given in Table 12. The AFM plot (Figure 29) shows the igneous affinities of these rocks, their strong Fe enrichment being similar to that of many tholeiitic gabbros (Figure 30). In the large metagabbro in the River Doe (Figure 23) the banded rock (B) is the most evolved geochemically, and the metadolerite (D) the least evolved (Figure 31). The latter has strong affinities with the other analysed metagabbros in the area. The hornblende-schists sampled show a range in composition and a complex differentiation history similar to that of the metagabbros. The metabasites as a whole are comparable chemically with metabasites elsewhere in the Loch Eil and Glenfinnan divisions, falling on the same trend lines, but with generally higher P_2O_5 and lower MgO (Rock et al., 1985).

SIX

Caledonian and post-Caledonian igneous rocks

INTRODUCTION

Intrusive igneous rocks belonging to several distinct suites are present in the Glen Affric district, most being representatives of suites present in adjacent areas. The age relations of these rocks are firmly established on evidence either from within the district or from neighbouring areas and are shown below:

Age	Rock type/suite	Age relations
Permo-Carboniferous	Camptonite, monchiquite	Cut Ratagain Plutonic Complex and Strathconon Fault within area of Sheet 72W (Kintail)
Lower Devonian	Felsite, lamprophyre	Cut Ratagain Plutonic Complex, but displaced by Strathconon Fault
Late- to post-tectonic (late Caledonian)	Glen Garry vein complex	Cut microdiorite-appinite suite, but possibly cut by latest members of this suite
	Microdiorite Suite: felsic porphyrite, microdiorite, appinite	Cut Cluanie Granodiorite
	Cluanie Granodiorite	Cut by felsic porphyrite and microdiorite
	Early felsic porphyrite	Cut by Cluanie Granodiorite
	Regional pegmatite and aplite	Cut by felsic porphyrite and microdiorite

In addition, one west-north-west-trending dyke on the south side of Glen Moriston, at [248 107], is of dolerite and basalt identical to that of the Tertiary igneous rocks of western Scotland.

REGIONAL PEGMATITE AND APLITE

Veins of pegmatite and aplite are distributed across most of the sheet, usually sparingly, but locally in greater numbers (Figure 32). They are composed chiefly of quartz, perthitic microcline and acid plagioclase. Muscovite and biotite occur in most of the pegmatites and aplites, and garnet, apatite and tourmaline are locally present as accessories. Individual crystals of feldspar vary from about 10 cm long in coarse pegmatite to a few millimetres in associated aplitic veins; both extremes are sometimes found within the same body. The form of the veins varies from irregular dykes intruded along joints (A and A' on Figure 32) to ramifying networks with poorly defined margins against the country rock (B and B' on Figure 32). In some cases the mineralogy of the pegmatites seems to reflect that of the country rock, with a greater concentration of mica in pegmatites which intrude pelite, although such a relationship is not usually apparent.

The minimum relative age of intrusion of the regional pegmatites is determined by the microdiorite suite, members of which cut the pegmatites. No pegmatites occur in the Cluanie Granodiorite. The maximum age is less well defined. Pegmatite dykes trending between north-east and south-east on the south side of Loch Beinn a'Mheadhoin (A on Figure 32) are partly recrystallised and are foliated in places approximately parallel to their margins. Veins of pegmatite folded by late minor (D3) folds are present at several localities (Plate 4) and locally occur in the axial planes of such folds [124 213]. South-east of Creag a'Mhain [097 070] concordant veins of pegmatite are apparently folded by late D3 folds and offshoots lie parallel to the axial planes of such folds (Figure 33). Although these veins seem to be closely associated with the folding the relationship remains in doubt in the absence of axial-plane schistosity in the pegmatite. Moreover, in this general area the regional pegmatites are known to cut pelite with crenulation cleavages associated with the late folds.

More than one generation of pegmatite may be present. On the north side of Loch Cluanie [118 105] quartz veins are cut by veins of pink pegmatite which in turn are cut by white granular pegmatite. The opposite relationship may also be seen (Plate 5). Elsewhere the regional pegmatites cut ptygmatically folded quartz veins and the pegmatitic intercalations in the pelites.

EARLY FELSIC PORPHYRITE

These are leucocratic to mesocratic rocks distinguished in hand specimen from the later felsic porphyrites by the larger size and greater frequency of the feldspar phenocrysts. Of the 46 occurrences mapped within the district (Figure 34) the majority are in the form of dykes and sheets trending

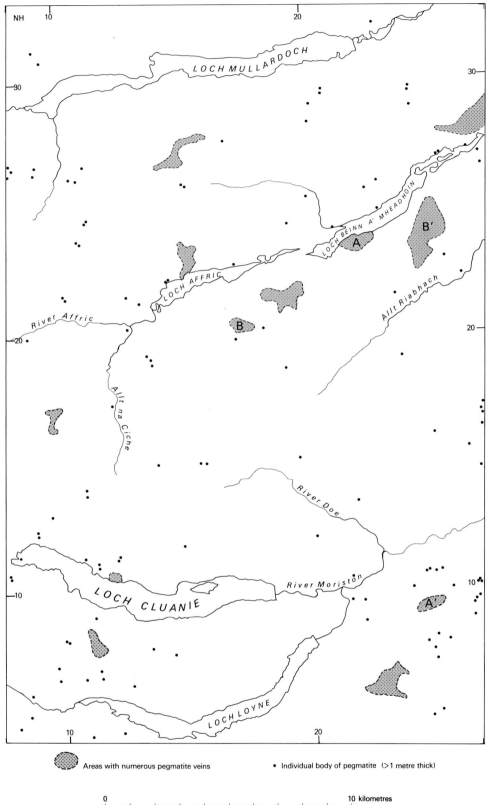

Figure 32 Distribution of regional pegmatite. For letters A, A′, and B, B′ see text.

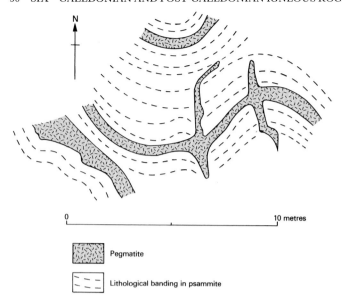

N

0   10 metres

▨ Pegmatite

▨ Lithological banding in psammite

Figure 33 Plan of pegmatite cutting Moine psammite at [097 070] 1.3 km S of Craig a'Mhaim.

between north-east and south-east. Others occur as small bosses or irregular masses a few metres across and some tens of metres long. The dykes and sheets are generally about 2 m thick but examples more than 5 m thick have been seen.

The field relationships of the early felsic porphyrites are firmly established. No member of the group cuts the Cluanie Granodiorite but they are spatially associated with it. At the margins of this major intrusion, however, sheets of early felsic porphyrite are cut by pegmatite and aplite veins extending from part of the Cluanie Granodiorite; for instance, by the roadside at [189 101] and in the Allt Coire Lair [127 109]. Blocks of early felsic porphyrite are found enclosed in microdiorite matrix in the Ceannacroc breccias (G on Figure 43 and p.58) and in a sheet of microdiorite in Coire Dho [193 148]. There is thus no doubt that within the Glen Affric sheet the early felsic porphyrites predate both the Cluanie Granodiorite and the microdiorites, but the close association with the former suggests a genetic relationship (Smith, 1979).

Petrographically the early felsic porphyrites are porphyritic microgranodiorites. They are commonly weakly schistose, and in thin section show some degree of recrystallisation. The phenocrysts, which are only slightly affected by

Plate 5 Micaceous psammite with concordant vein of white pegmatite cut by pink pegmatite. North side of Loch Cluanie [121 106]. D871.

Cluanie Granodiorite   •   Felsic porphyrites

Figure 34   Distribution of early felsic porphyrites in relation to the Cluanie Granodiorite.

this recrystallisation, are dominantly oscillatory zoned albite-oligoclase up to 5 mm across accompanied by smaller and subsidiary green hornblende and biotite. The latter minerals occur also in the granoblastic groundmass which is composed mainly of plagioclase with subsidiary quartz. Sphene is a common accessory mineral.

CLUANIE GRANODIORITE

This major igneous body, which lies at the east end of Loch Cluanie, is some 7 km long in an east–west direction and about 4 km wide (Leedal, 1953). It is formed chiefly of pale grey to pale pink hornblende-granodiorite and has been subdivided according to the presence and abundance of megacrysts of potash feldspar (Figure 35). Such subdivisions have locally sharp, but commonly transitional, boundaries and there is no evidence to suggest that they reflect distinct intrusive episodes. The hornblende-granodiorite without megacrysts on the east and south margins corresponds in part to a more basic zone described by Leedal (1953, p.41). The small area with conspicuous biotites on the east margin (Figure 35) may be a chilled variety of the hornblende-granodiorite without megacrysts. At the east margin of the intrusion the granodiorite and neighbouring country rock are intensely veined by aplite and local pegmatite (Figure 35). Aplites and pegmatites locally occur at the contacts of the intrusion elsewhere, forming a zone of veining varying from a few centimetres to a few metres in thickness. They also form isolated veins in the adjoining country rock. The aplitic rocks generally cut both the granodiorite and the country rock, but in places pass into nonporphyritic hornblende-granodiorite and may even be cut by it, suggesting that their intrusion was closely linked with that of the main part of the complex. The body of almost pure vein quartz which has been mapped at a locality [188 096] on the east side of the intrusion is probably unrelated to it.

Where the outer contact of the complex is well exposed it can be seen to dip at various angles, chiefly outwards, but inwards at one locality on the southern side. Detailed mapping has shown that the attitude of the contact is usually discordant to the lithological banding in the country rock and not concordant as suggested by Leedal (1953). No support has been found for Leedal's view that the regional strike and dip of the country rocks is partly the result of magma pressure. In places, rafts of country rock in the marginal aplite seem to have retained their original attitude despite the intricate boundaries which are apparently due to stoping. On Creag na Mairt [175 115] the roof of the intrusion is seen with pendant blocks of psammite. Other possible roof pendants occur on Creag nam Peathrain (Leedal, 1953) and at localities [184 112] and [183 110] (A on Figure 35). Contact metamorphism of the country rock by the granodiorite is slight, although Leedal (1953, p.52) reported the formation of metasomatic potash feldspar, plagioclase, hornblende, sphene, zircon and allanite in psammite xenoliths.

Igneous fabrics (flow foliation) have been recorded at only two localities, north of Loch Cluanie (Figure 35), the strike in each case being approximately parallel to the margin of the intrusion. Jointing is well developed (Leedal, 1953, p.41), with one prominent set striking between north-north-

west and north-north-east the other between east-north-east and east. According to measurements carried out during the present survey, the latter set is most prominent, and dip about 40° north. The former is associated with small faults and crush zones along which dykes of the microdiorite suite have been sinistrally displaced for distances of up to 1.5 m (Leedal, 1953). Sheet joints parallel to the topography are developed to the south of Loch Cluanie.

The petrography of the complex has been described in detail by Leedal (1953). The hornblende-granodiorite without megacrysts has an average grain size of 1–2 mm, though close to the outer contact this is reduced to 0.5–0.8 mm. It consists of 15–20% quartz, 10–15% potash feldspar, 50–60% plagioclase, 2–3% biotite, 5–8% hornblende, 2% magnetite, 1–2% sphene and 1% apatite. The plagioclase crystals are up to 3 mm in diameter, subrectangular in section, though very irregular at the edges. Oscillatory zoning from An_{30} to An_{15} is well developed, with concentric shells 0.3–0.5 mm thick of normally zoned plagioclase bounded by a sharp reversion to An_{30}. The potash feldspar forms crystals up to 2 mm, but generally less than 1 mm, which corrode and replace the outer parts of the plagioclase. Myrmekite is abundantly developed in plagioclase crystals within 0.5–1 mm of their boundaries with potash feldspar. The potash feldspar is orthoclase which contains little or no perthite, and microcline twinning is absent. Hornblende forms prismatic crystals up to 1.5 mm long, with α—pale yellow, β—olive green, γ—blue-green. Dark umber brown biotite is subordinate to hornblende and forms hexagonal books up to 2 mm in size. In many places it is chloritised along the edges. Sphene forms euhedral crystals up to 0.5 mm in size, apatite crystals up to 0.5 mm are irregular with rounded ends, and magnetite forms euhedra, which are often icositetrahedra rather than octahedra.

The hornblende-granodiorite with conspicuous biotite is a variant of this rock in which biotite crystals up to 5 mm across form approximately 5% of the rock. The biotites are randomly oriented, and the rock also contains quartz crystals up to 5 mm in size which are possible xenocrystic.

The megacrystic varieties of the granodiorite are slightly more acid in composition and coarser-grained than the non-megacrystic varieties. The grain size is 3–4 mm, and an approximate mode is 20% quartz, 20% potash feldspar, 50% plagioclase, 4–5% hornblende, 2% biotite, 1–2% magnetite, 1–2% sphene, 1% apatite and 0.5–1% zircon. The plagioclase crystals are very similar to those of the non-megacrystic granodiorite, but the quartz and potash feldspar crystals are larger, mostly reaching 2–3 mm, while a proportion of the potash feldspar crystals reach 10–15 mm (megacrysts). These megacrysts are roughly euhedral in hand specimen, but in thin section are seen to have very irregular margins and to enclose numerous 0.2–1 mm euhedral crystals of plagioclase, hornblende and sphene. Perthite occurs as stringers and irregular patches. Microcline twinning is poorly developed, but more conspicuous than in the non-megacrystic granodiorite. The plagioclases external to the potash feldspar crystals are corroded and replaced, and develop myrmekite in the same way as in the non-megacrystic granodiorites.

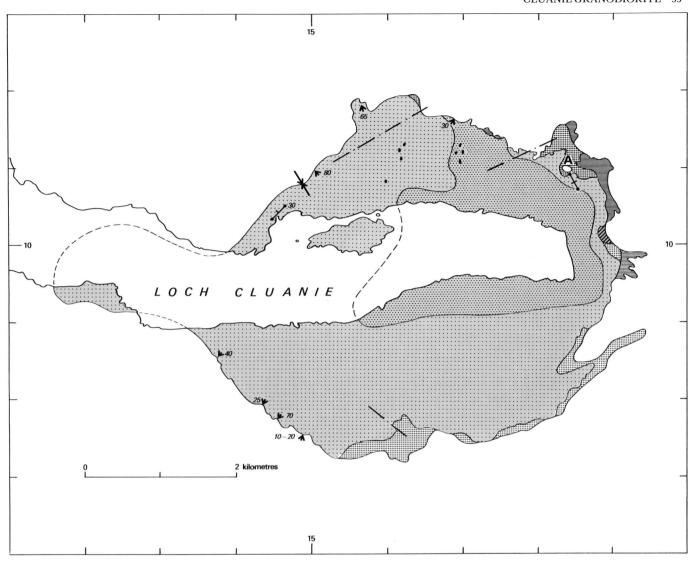

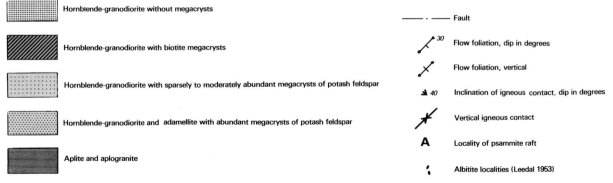

Hornblende-granodiorite without megacrysts

Hornblende-granodiorite with biotite megacrysts

Hornblende-granodiorite with sparsely to moderately abundant megacrysts of potash feldspar

Hornblende-granodiorite and adamellite with abundant megacrysts of potash feldspar

Aplite and aplogranite

Geological boundary

Fault

Flow foliation, dip in degrees

Flow foliation, vertical

Inclination of igneous contact, dip in degrees

Vertical igneous contact

A — Locality of psammite raft

Albitite localities (Leedal 1953)

Figure 35 Map of Cluanie Granodiorite.

The more and less abundantly megacrystic granodiorites are otherwise identical, and the mapped boundary between them is gradational.

The aplites which intrude the margins of the pluton are of monzogranite composition (30% quartz, 30% potash feldspar, 40% plagioclase). The grain size is 0.2–1 mm. They contain 0.5–1% of euhedral magnetite and approximately 0.5% of pale pink amoeboid garnet crystals up to 0.5 mm in size. The texture is largely saccharoidal, but in places slight shearing is apparent.

Approximate ages of intrusion of 417 Ma and 425 Ma have been suggested for the Cluanie Granodiorite (Pidgeon and Aftallion, 1978, p.192; Brook in Powell, 1983, p.294). There is, however, a discrepancy between this figure and the reported ages for other Late Caledonian granitic intrusions in the West Highlands. The Strontian Granite, for instance, is dated at 435 Ma (Powell, 1983 p.294) yet is known to postdate the microdiorites, whereas the Cluanie Granodiorite predates them.

DYKE-LIKE BODY OF GRANODIORITE

A small granodiorite body which crops out at the west edge of the map is well exposed in the Allt a'Chaoruinn Bhig, 700 m north of Cluanie Inn [076 127], and extends westwards, across strike, for about a kilometre into Sheet 72W (Kintail). The rock is a pale pink granodiorite with very subsidiary biotite and hornblende. In the stream the southern edge of the intrusion is bounded by veined schist about 100 m wide in contrast to the northern edge along which the junction is sharp with no disruption of the adjacent country rock. The granodiorite is also exposed in a gully to the west of the main stream where it forms a vertical dyke-like body about 30 m wide, which appears to thin and die out westwards. Though the granodiorite can probably be classed with the Caledonian igneous rocks its relationship to the other suites of this age is not known.

MICRODIORITE SUITE

General

The great majority of the minor intrusions mapped on the sheet belong to this suite. For the purposes of field classification the suite has been subdivided as follows:

Coarse-grained	*Medium- to fine-grained*
Hornblendite and biotitite Hornblende-diorite Granodiorite } Appinite	Melamicrodiorite Microdiorite Leucomicrodiorite Felsic porphyrite } Microdiorite

The more numerous finer-grained members range from leucocratic felsic porphyrite to melamicrodiorite which are the end members of what is essentially a continuum from acid to basic (Smith, 1979). The boundary between microdiorite and melamicrodiorite in this classification is taken as 50% mafic minerals and that between microdiorite and leucomicrodiorite as 25% mafic minerals. The appinites are principally composed of mafic hornblende-diorite, but include acid to ultrabasic varieties, all of which may form part of a single intrusion. Transitions from appinite to microdiorite are known and melamicrodiorite net-veined by leucomicrodiorite is not uncommon.

The field relationships of the suite indicate that it postdates the Cluanie Granodiorite (Figure 36) and xenoliths of the early felsic porphyrites are caught up in microdiorite at [193 148] as well as in the Ceannacroc breccias (p.58). Both in this district and elsewhere felsic porphyrite is known to cut microdiorite, but not vice-versa (Smith, 1979), showing that the more acid members generally postdate the more basic. The relative ages of the microdiorite suite and the Glen Garry vein complex are not entirely clear because of similarity of rock type (Fettes and McDonald, 1978, p.346). The vein complex rocks are known to cut microdiorites and are themselves cut by felsic porphyrite (Fettes and MacDonald, 1978, p.338). Such felsic porphyrites could belong to the vein complex, the microdiorite suite, or to neither of these. Members of both the microdiorite suite and the vein complex occur as matrix in bosses of breccia immediately north of the mouth of the River Loyne (Figure 43). These, the Ceannacroc breccias, are discussed separately below.

Appinite

Some 40 appinite bodies have been found on the sheet (Figure 36), their distribution approximating to that of the microdiorites. They tend to occur as sheets up to 5 m thick and as small bosses (several tens of metres across), the latter comprising a wide variety of rock types within a single intrusion and locally including fragments of country rock. Examples of such bosses can be seen on the hillside [185 127] north of the east end of Loch Cluanie.

In the more basic appinites, the texture is dominated by interlocking crystals (2 to 4 mm) of hornblende with subsidiary biotite and plagioclase, but in more feldspar-rich varieties the mafic minerals tend to be included in poikilitic oligoclase or andesine. The poikilitic texture is also seen in some granodioritic rocks of the appinite suite in which the mafic mineral is chiefly biotite and in which the plagioclase is accompanied by subsidiary potash feldspar. Four pyroxene bearing appinites have been recorded (Figure 36), these being the northernmost examples of a number of such intrusions more common in the area south of Loch Garry (Sheet 62E, Loch Lochy). Some appinites show the effects of recrystallisation along grain boundaries, but a well-developed schistosity occurs only on the margins of the intrusions where it is subparallel to the contact.

Microdiorite

Within the Glen Affric district the microdiorites for the most part occur as intrusive parallel-sided sheets, but north-east of

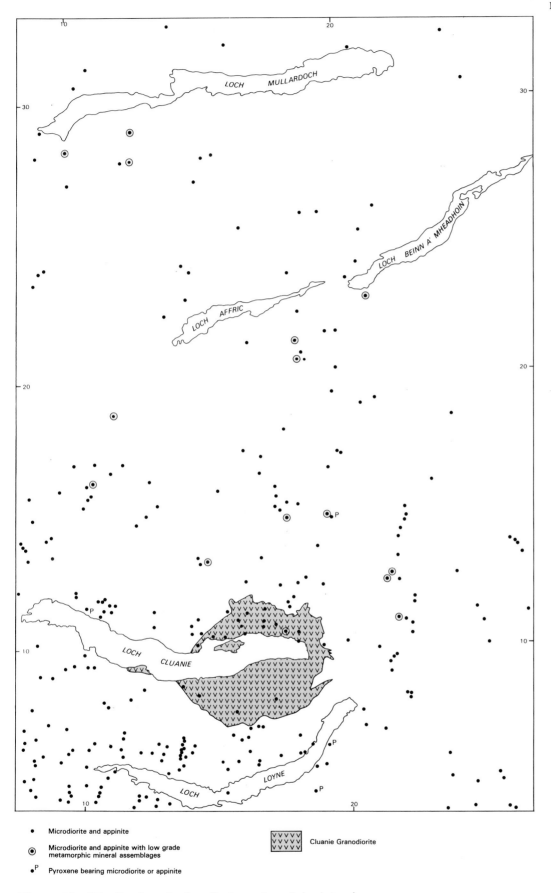

- • Microdiorite and appinite
- ⊙ Microdiorite and appinite with low grade metamorphic mineral assemblages
- •P Pyroxene bearing microdiorite or appinite

Cluanie Granodiorite

Figure 36   Distribution of microdiorite and appinite intrusions.

Loch Cluanie there are a number of irregular intrusions associated with the Ceannacroc breccias (p.58). Over 200 microdiorite sheets have been mapped (Figure 36), the vast majority being in the southern half of the district where there are, on average, 2 to 5 intrusions per square kilometre. Most of them are cross-cutting and average one metre in thickness with a range from several metres down to a few millimetres. From Figure 37 it can be seen that most intrusions strike east-north-east and dip about 45° south-south-east. This orientation is similar to that of the suite as a whole (Smith, 1979). A number of the sheets display an internal schistosity which Smith (1979) has related to westward overthrusting in the case of the eastward-dipping sheets and eastward overthrusting for those inclined to the west. An example of the former can be seen at [078 073] 1 km west-south-west of Craig a'Mhaim and of the latter at [083 150] 1 km west of A'Chralaig. No examples have so far been described of microdiorite sheets folded together with Moine country rock as reported from the Glenfinnan and Loch Sunart areas by Talbot (1983).

The unmetamorphosed microdiorites typically consist of subhedral tablets of hornblende and zoned andesine (either or both of which may occur as phenocrysts) together with biotite and subordinate interstitial quartz and potash feldspar. In some intrusions the feldspar forms irregular interlocking plates which are poikilitic towards the mafic minerals, a texture typical of the appinites. In others a plumose texture has been noted in the feldspar. A single pyroxene-bearing example has been found, on the north side of the valley of the River Doe (Figure 36).

Most of the microdiorites on the sheet are metamorphosed to some degree, varying from incipient recrystallisation along grain boundaries to complete reconstitution. The recrystallisation is commonly accompanied by the formation of a fabric in the mafic minerals, which gives the rock a schistosity and/or a lineation. The brown hornblende is partly or entirely replaced by biotite and actinolite in many intrusions and in a few cases the original igneous minerals have been replaced by a low-grade metamorphic assemblage of albite, quartz, biotite, epidote, actinolite and sphene with varying amounts of calcite (Smith, 1979).

Isolated examples with low-grade mineral assemblages occur north of Loch Cluanie (Figure 36), but are absent further south. This is in general agreement with the southward increase in metamorphic grade noted by Smith (1979) in this part of the Highlands. However, the overlap of occurrences of intrusions showing different degrees of metamorphism supports the view that the microdiorites were either intruded over a period of time while the rock temperature was declining or that there was more than one period of intrusion. These conclusions need to be tested by a detailed study of their structural relationships and geochemistry.

Felsic porphyrite

There are about 100 intrusive sheets of felsic porphyrite known in the district. Their main distribution is in a belt from the south-west corner of the map, tapering north-eastwards to the east end of Loch Beinn a'Mheadhoin (Figure 38). A plot of the orientation of 54 sheets from the southern half of the map area (Figure 39) indicates that the attitude of the sheets is similar to that of the microdiorites, with the majority trending east-north-east.

Petrographically the felsic porphyrites are porphyritic microgranodiorites or quartz diorites in which phenocrysts 2 to 4 mm across of oscillatory-zoned albite-oligoclase are accompanied by generally smaller and less-numerous phenocrysts of biotite and hornblende. The groundmass in the more-acid examples is granular quartz, acid plagioclase and potash feldspar in roughly equal proportions accompanied by hornblende and biotite. The less-acid examples grade into leucomicrodiorite by a reduction in the proportion of plagioclase phenocrysts. Many of the felsic porphyrites are more or less recrystallised with the development of similar mineral assemblages to those found in the metamorphosed microdiorites (i.e. alb-qtz-biot-epid-act-sph-(calc)) and a schistosity of aligned biotites and/or hornblendes which wrap around the rotated remnants of zoned plagioclase phenocrysts. Garnet has not been noted in metamorphosed felsic porphyrites in the Glen Affric district, but has been reported from nearby in the Loch Lochy (Sheet 62E) area (Smith, 1979, Figure 2). Unmodified and recrystallised felsic porphyrites with hornblende, actinolite and biotite are found throughout the belt, but examples with low-grade mineral assemblages have been found only north of Loch Cluanie (Figure 38). This distribution is similar to that of the microdiorites. However, the felsic porphyrite intrusions

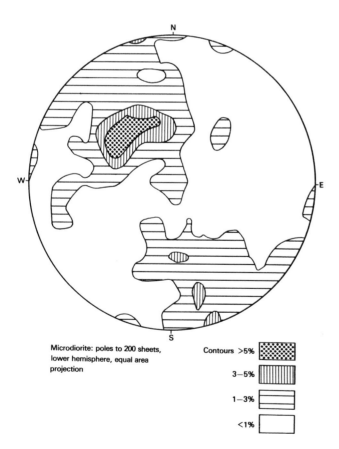

Microdiorite: poles to 200 sheets, lower hemisphere, equal area projection

Contours >5%

3–5%

1–3%

<1%

Figure 37 Stereoplot of poles to microdiorite intrusive sheets.

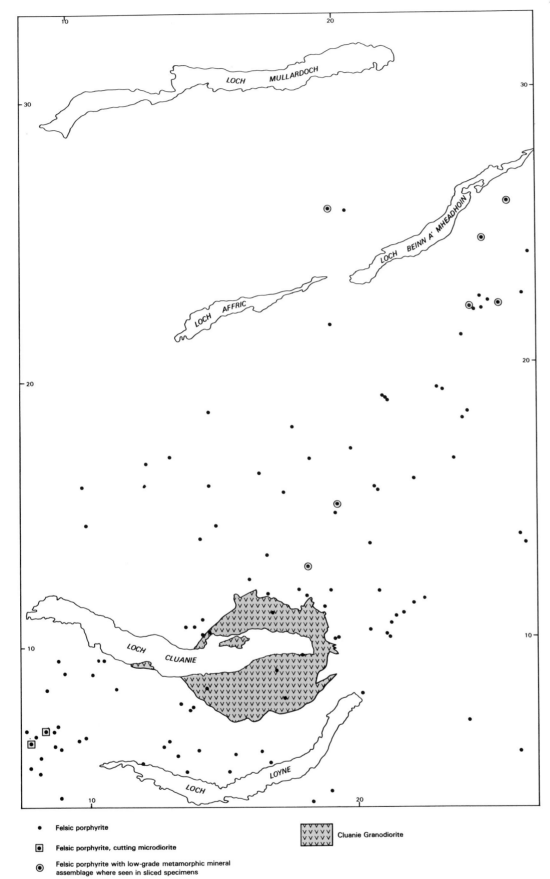

Figure 38   Distribution of felsic porphyrite intrusions.

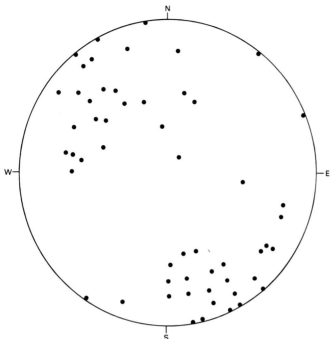

Felsic porphyrite: poles to 54 sheets, lower hemisphere, equal area projection

Figure 39 Stereoplot of poles to felsic porphyrite intrusive sheets.

which are known to postdate the vein complex rocks are unmetamorphosed.

GLEN GARRY VEIN COMPLEX

This term is applied to a suite of unmetamorphosed minor intrusions composed predominantly of granodiorite with minor quartz diorite, tonalite and granite which occur as irregular bodies and veins ramifying through the country rock (Fettes and MacDonald, 1978). These rocks are also found in the matrix of breccia bosses adjacent to the northern edge of their main area of distribution, including the matrix of the Ceannacroc breccias discussed below. The northern limit of the vein complex in the southern half of the Glen Affric district is quite sharply defined (Figure 40), but scattered veins probably related to the complex have been recorded up to 10 km beyond this limit. Veins of the complex cut the schistose members of the microdiorite-appinite suite as well as non-schistose diorites which have a well-preserved igneous mineralogy and texture. Fettes and MacDonald (1978) suggest that the latter should be classed with the microdiorite-appinite suite rather than with the vein complex rocks on the basis of chemistry and percentage of modal minerals, but their status remains in doubt.

The description of the Glen Garry vein complex by Fettes and MacDonald (1978, pp.339–341) applies to the part of the complex which lies within the Glen Affric district. The veins have sharp, cross-cutting margins and are massive without markedly oriented internal fabrics. Displacement of

banding in the country rock and older veins is common, indicating dilation. Within the Glen Affric district the major dyke-like bodies have a general east-north-east trend and the larger masses tend to be elongated in the same direction (Figure 41). This is more or less parallel to the trend of members of the microdiorite suite in the area. Fettes and MacDonald note that there is no meaningful pattern in the spatial variation of the different rock types, but that, temporally, the more acid varieties were latest in the intrusion sequence. Many of the intrusions carry xenoliths of the country rock and of earlier emplaced igneous rocks.

The mineralogy of the vein complex rocks is summarised in Figure 42, and the following account is slightly modified from Fettes and MacDonald (1978, p.342). The rocks are medium to coarse grained and (if the dioritic rocks are excluded) characteristically nonporphyritic. Plagioclase, quartz, potash feldspar, biotite and hornblende are the main constituents. Accessory minerals, which are commoner in the more basic rocks, include sphene, apatite, epidote, carbonates, allanite, zircon and ore minerals; in addition there are abundant secondary micas and chlorite. The hornblende forms subhedral grains and is commonly marked by secondary alteration. Euhedral crystals of hornblende are enclosed in plagioclase and sphene. Biotite is generally anhedral although in some rocks it projects into the outer zones of plagioclase crystals. Quartz forms anhedral grains. Plagioclase is typically subhedral, often with poikilitic margins and ranges in size up to 4 mm. The crystals are characteristically zoned with turbid centres; the cores vary in composition from An_{26} to An_{12} and the rims are more sodic. Myrmekitic intergrowths are common. The potash feldspar is generally confined to small interstitial grains (0.1 to 0.8 mm) although in the more acid rocks it may form large poikilitic patches; some of the larger crystals show perthitic intergrowths.

From a consideration of the chemistry and mineralogy Fettes and MacDonald (1978) suggest that the complex evolved from a quartz diorite magma under relatively high water pressures and that intrusion may have been related to late Caledonian stresses associated with faulting on the Great Glen Fault.

CEANNACROC BRECCIAS

This name is applied to a number of breccia bodies which are chiefly found within a radius of 4 km of Ceannacroc Lodge [227 113], most notably on the hillsides just north of the mouth of the River Loyne (Figures 40 and 43). They form rounded, ice-moulded knolls which stand above the general level of their surroundings. The dimensions vary from a few metres to about 500 m in length and up to 100 m across.

The clasts in the breccias are randomly oriented, angular, subrounded and less commonly well rounded (Plate 6). They average about 0.3 m in length, but range from about 1 m to very small fragments which locally form the matrix. From Table 13 it can be seen that all the clasts are of the local country rock. They are not, however, always formed of the rock cropping out in the immediate neighbourhood and where, for instance, there are fragments of granitic gneiss in metasediment, a distance of transport of up to several hundred

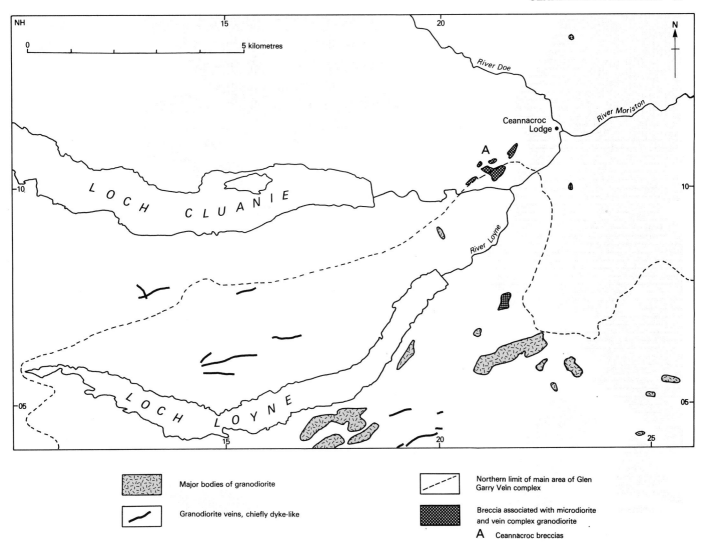

Figure 40 Distribution of Glen Garry vein complex rocks and of breccias associated with microdiorite and vein complex granodiorite.

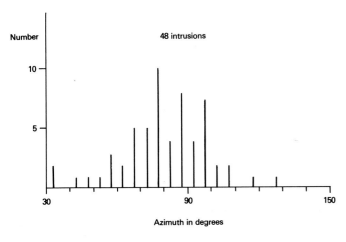

Figure 41 Trend of minor intrusions of the Glen Garry vein complex in the area between Loch Loyne and Loch Cluanie.

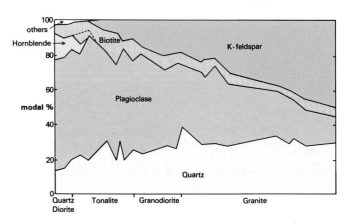

Figure 42 Modal variation in different rock types of the Glen Garry vein complex.

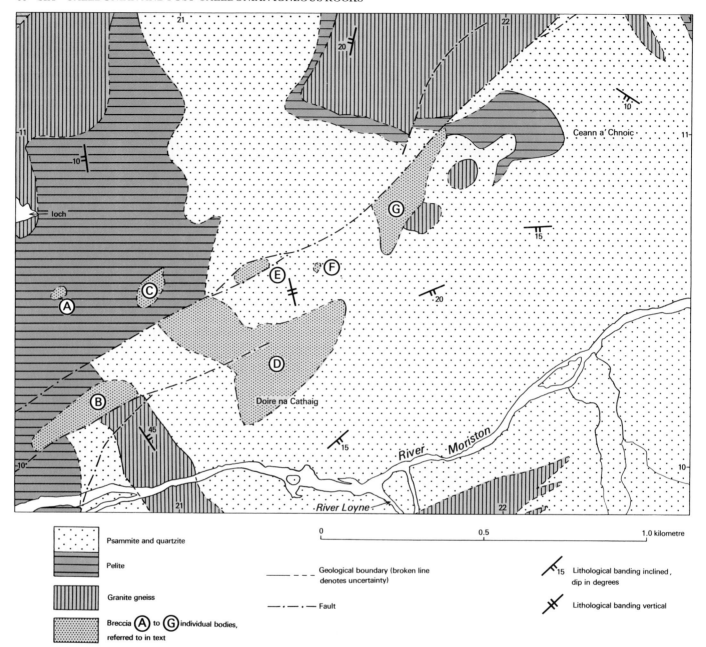

Figure 43   Simplified geological map of area north of the mouth of the River Loyne.

metres is indicated. Generally the clasts are more fractured and weathered than the same rocks in situ.

The igneous rocks associated with the breccias are microdiorite, felsic porphyrite and vein complex granodiorite. Some relationships are as shown in Table 14.

In thin section it can be seen that the contacts of schistose microdiorite with the fragments is sharp and that the schistosity in the microdiorite is confined to this rock. There is no evidence of thermal metamorphism associated with felsic porphyrite, microdiorite or vein complex granodiorite.

A maximum limit for the relative age of the breccias is determined by the presence of clasts of early felsic porphyrite and of aplite identical to that forming part of the Cluanie Granodiorite. A minimum age limit is provided by the pres-

ence of vein complex granitic rocks, forming or cutting the matrix.

The form of the breccia bodies in plan suggests that they are pipe-like with a tendency to be elongated in a northeasterly direction. This suggests the possibility that they could be related to the faults of this trend which are spatially associated with them, but no evidence is forthcoming to support this idea. Though the clasts do not seem to be far travelled, some have been rounded, presumably during transport. The presence of very small fragments forming the matrix in places, and the greater degree of weathering and fracturing of the clasts compared with the same rocks outwith the breccias, suggests the involvement of gas streaming rather than passive stoping during the passage of the magma.

Table 13 Nature of clasts in Ceannacroc breccias

Breccia mass	Country rock	Clasts
A	Pelite, partly, brecciated	Pelite, granitic gneiss
B	Granitic gneiss, quartzite	Granitic gneiss, quartzite
C	Psammite, pelite, hornblende-schist	Chiefly granitic gneiss
D	Psammite and quartzite	Chiefly granitic gneiss, amphibolite, pegmatite, pelite, quartzite, aplite
E	Not known	Chiefly granitic geneiss, quartzite
F	Psammite and quatzite	Psammite, quartzite, hornblende-schist
G	Quartzite, granitic gneiss	Granitic gneiss, pelite, hornblende-schist, pegmatite, early felsic porphyrite
H [232 135]	Pelite, psammite	Chiefly pelite, quartzite, psammite, granitic gneiss, amphibolite

For localities A to G see Figure 43

The inter-relationships of the associated igneous rocks suggest more than one period of intrusion, with microdiorite and felsic porphyrite cut by granitic rocks of the vein complex. All these rock types locally form the matrix of the breccias and certainly vein them. The relationships recorded in breccia mass B of Table 14, where xenoliths of schistose microdiorite are enclosed in felsic porphyrite, but apparently also form the matrix, must be treated with reservation since the xenoliths could be a slightly more basic variety of early felsic porphyrite, easily confused with microdiorite.

The Ceannacroc breccias resemble the explosion breccias of the Kentallen district (Bowes and Wright, 1967) both in size and shape, but differ from them in that there is little diversity in the associated igneous rocks and no undoubted pneumatolytic alteration. Though the pipes and clasts may have formed initially as a result of explosive activity conse-

Table 14 Field relationships of Ceannacroc breccias

Breccia mass and location within outcrop (Figures 40 and 43)		Field relationships
A	General	Matrix of felsic porphyrite and granodiorite
B	Centre of mass	Late, little-metamorphosed, felsic porphyrite body. 2 m seen, with xenoliths of schistose partly recrystallised leucomicrodiorite which apparently locally forms breccia matrix
	West margin	Vein complex granodiorite forms agmatite
F	West margin	Matrix locally of vein complex granodiorite
G	South end	Matrix of schistose microdiorite which locally occurs as discrete bodies with a north-trending schistosity. Microdiorite cut by microgranite veins.
	West side	Matrix of leucomicrodiorite. Fragments include early felsic porphyrite
	North end	Approximately circular mass of schistose low grade microdiorite 2 m across which intricately veins the surrounding breccia. The mass contains scattered xenoliths of quartzite. About 2 m to SW a similar discrete microdiorite mass is separated from the breccia by a zone of agmatite about 10 cm thick.
	General	Part of matrix seems to be comminuted rock

Plate 6 Breccia associated with late Caledonian minor intrusions. Fragments of granitic gneiss, hornblende-schist and psammite. Matrix of comminuted material. Glen Moriston west of Ceannacroc Bridge [215 105]. D866.

quent on the release of gas pressure (Bowes et al., 1963), the subsequent transportation of the fragments could have been by a liquid rather than a gaseous phase, with a filter press mechanism resulting in local concentration of clasts without an obvious igneous matrix—a feature characteristic of several of the breccias.

FELSITE

A few dykes of pink felsite occur in the west of the Glen Affric district. Three dykes, each a few metres wide, in the north-west corner of the map trend east–west (Figure 44) whereas the three exposures recorded south-west of Loch Affric belong to a single south-east-trending dyke. The latter, which is about 8 m wide at one locality, is the most easterly member of a dyke swarm on the adjacent Kintail (72W) Sheet which is displaced by the Strathconon Fault but known to cut the Ratagain Plutonic Complex; it is taken to be of Lower Devonian age (see also the discussion of the lamprophyres below).

Petrographically the felsites are fine-grained quartzofeldspathic rocks with a few flakes of muscovite and a little opaque iron oxide. There are rare phenocrysts of potash feldspar and sericitised plagioclase up to 2.5 mm across. Muscovite-sphene-oxide pseudomorphs after biotite may also be present.

MINETTE SUITE

Rocks of the minette suite occur as dykes, locally up to 6 m thick; they are part of a swarm occurring widely in the Northern Highlands (Smith, 1979). For the most part they are conspicuously red in colour, but a few are dark grey. They are found chiefly in the western part of the Glen Affric district (Figure 45), the majority trending between 90 and 140° (Figure 46). Within the Glen Affric district they postdate the Cluanie Granodiorite. Pebbles of minette suite rocks occur in basal Devonian strata in Strath Rannoch (Peach et al., 1912, p.113) and two sills possibly belonging to the suite are said to cut basal Devonian breccia near Loch

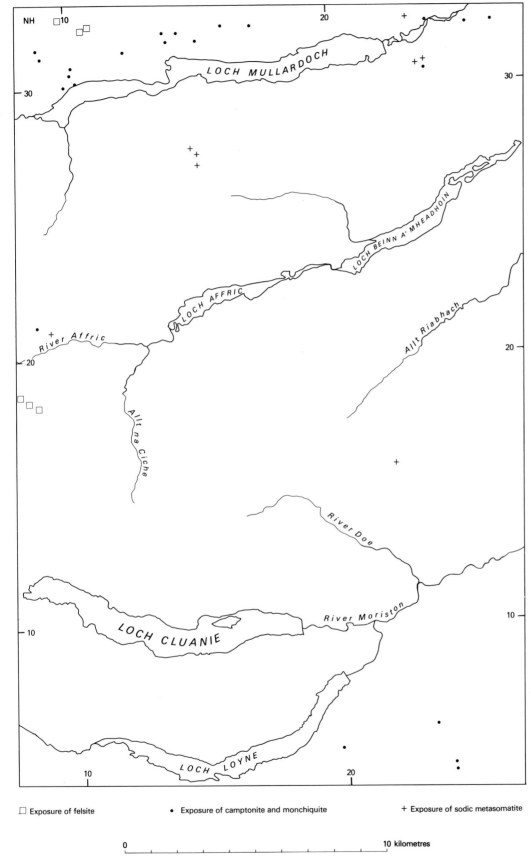

☐ Exposure of felsite   • Exposure of camptonite and monchiquite   + Exposure of sodic metasomatite

0   10 kilometres

Figure 44   Exposures of felsite, camptonite, monchiquite and sodic metasomatite.

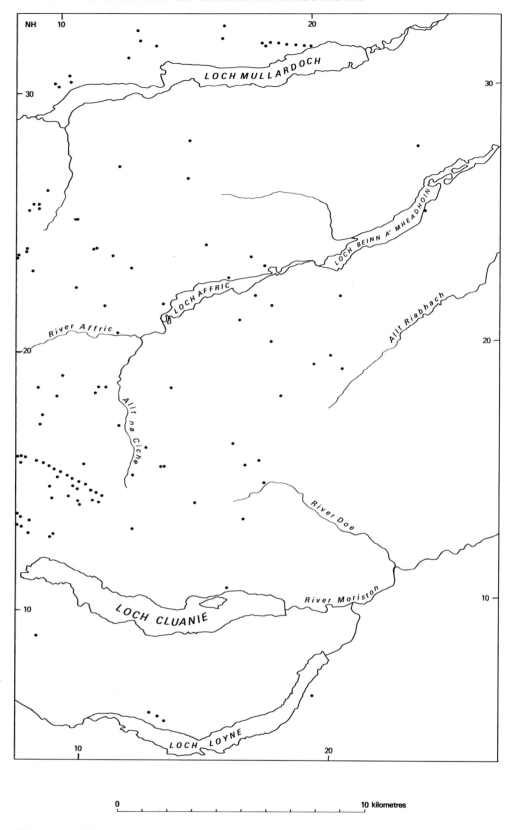

Figure 45 Exposures of minette suite dykes.

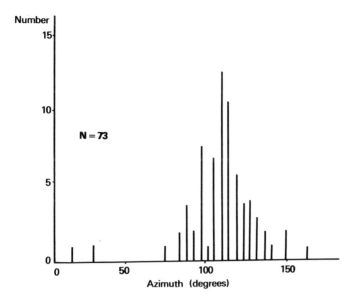

Figure 46 Trend of minette suite dykes.

Garve (Horne and Hinxman, 1914, p.53; Smith, 1979 p.685). In the area west of the Glen Affric district there are examples of lamprophyre cutting felsite and felsite cutting lamprophyre indicating that the periods of intrusion of these two suites overlapped.

Petrographically most of the dykes are pyroxene minettes, rich in corroded colour-zoned phenocrysts of biotite and colourless or pale green clinopyroxene. They are identical to those described by Read et al. (1926) from the Strath Oykell region to the north and classed by him as augite minettes.

Carbonate or serpentinous pseudomorphs after olivine phenocrysts are fairly common, as is pale brown hornblende. In a few cases hornblende is the dominant ferromagnesian mineral and the rocks are then classed as vogesites. In most specimens of both types the fine- or medium-grained groundmass is dominantly feldspathic. Orthoclase and acid plagioclase are usually both present but their turbid and partly decomposed nature makes it difficult to determine their relative abundance. In a few cases the dominant feldspar appears to be acid plagioclase and the suite may therefore include spessartites and kersantites. In National Grid square [08 23] there are several minette dykes with phenocrysts of acid plagioclase. Quartz occurs interstitially and as xenocrysts. Apatite and magnetite are abundant in most specimens.

CAMPTONITE-MONCHIQUITE SUITE

A few intrusions of this Permo-Carboniferous suite have been mapped, chiefly north of Loch Mullardoch (Figure 44), where they probably represent the southern edge of the Monar swarm to the north (Ramsay, 1955; Rock, 1982). They are mainly in the form of dykes up to 2 m across which trend between east-south-east and east-north-east but sheets concordant to the lithological banding have also been observed. Most of the intrusions are camptonites or camptonitic basalts containing small amounts of biotite and brown amphibole. The monchiquites, which are rich in olivine, pyroxene and brown amphibole are represented by only one dyke [198 054]. Another dyke [102 302], which is classed with the monchiquites on the published 1:50 000 sheet, is a pyroxene-rich ultramafic rock (yamaskite).

SEVEN

Metasomatic rocks

SODIC METASOMATITES WITH HORNBLENDE AND PYROXENE

Rocks of this type were originally described from Glen Cannich (Tanner and Tobisch, 1972) and several more occurrences are now known (Figure 44). They postdate the formation of minor folds in the country rock and on the adjacent Loch Quoich (62W) Sheet they are cut by microdiorite and felsic porphyrite of the microdiorite-appinite suite (Peacock, 1972). The metasomatites can be recognised in the field by bleaching of the country rock and the development of spots of dark ferromagnesian minerals. The contacts with the unmetasomatised rock are commonly gradational. Some crosscut the lithological banding, others are parallel to it. Fold structures may be preserved mimetically in metasomatised rock. Most of the bodies of metasomatite are small, but one outcrop on Creag na h-Iolare [236 306] extends at least 100 m in a north-north-west direction and is about 20 m wide (Tanner and Tobisch, 1972, p.161).

Petrographically the metasomatites are diverse, generally consisting of a variably textured groundmass of albite and quartz (up to 0.5 mm) with porphyroblasts of hastingsite or hastingsitic hornblende and diopside-hedenbergite with a variable Na_2O content.

Tanner and Tobisch (1972) have pointed out the similarity of the mineralogy of the metasomatites to that of fenite zones surrounding alkaline complexes, though they differ from fenites in their bulk chemistry. They suggested a genetic relationship to saturated syenite magmas, though no such bodies are present at the surface within the district. The metasomatites north-east and south-east of Loch Mullardoch (Figure 44) are immediately adjacent to east–west faults which could have acted as pathways for the metasomatising fluids, but no such relationship is known for the other occurrences.

METASOMATIC ALBITITE

Leedal (1953) reported the occurrence of a number of bodies of metasomatic albitite spatially associated with the Cluanie Granodiorite (Figure 35). These bodies range in width from 10 cm to about 2 m. They are related mainly, but not entirely, to the north-trending joints and faults within the granodiorite complex. In these rocks quartz is replaced by feldspar, plagioclase is albitised, biotite and hornblende largely disappear and there is an introduction of late-stage cleavelandite (lamellar albite). In some cases the microcline phenocrysts in the hornblende-granodiorite have grown to porphyroblasts several centimetres in length, and mariolitic cavities have been formed by the leaching of calcite. Bodies of albitite also occur in granodiorite dykes outwith the Cluanie Granodiorite in the Coire Beithe area [137 077]. These dykes are now classified with the Glen Garry vein complex, which is later than the intrusion of the Cluanie Granodiorite. This, and the association with the faulting which postdates the microdiorites, suggests that the albitisation, though spatially associated with the Cluanie Granodiorite, postdates both this body and the later intrusive igneous rocks of late Caledonian age. The relationship to the Lower Devonian lamprophyres, which also cut the granodiorite, is not known.

EIGHT

Faults and mineralisation

FAULTS

Most of the significant faults in the district trend between SE and NE, with the majority between ENE and NE (Figure 47). In the extreme NW of the district is the Strathconon Fault (A on Figure 47), which, as well as other less important faults, is dominated by lateral movement. On most faults, almost all of which are vertical or inclined at high angles, the sense of movement is much less clear. Many faults, including those which have little or no effect on the stratigraphy, are marked by crush zones and, in some cases, by deeply decomposed rock and carbonate.

There is only a very small section of the Strathconon Fault system in the Glen Affric area, here a single fault separating rocks of the Morar and Glenfinnan divisions. On the adjoining Kintail Sheet (72W) the net sinistral movement, as determined from the displacement of Lower Devonian dykes, is about 8 km. In addition, a considerable south-eastward downthrow has been deduced by Simony (1963) from a comparison of the fold structures on either side of the fault system.

South of Loch Beinn a Mheadhoin there are a number of ENE-trending faults (B on Figure 47), the most important of which has an apparent sinistral displacement of 0.6 km and passes eastwards on the adjoining Invermoriston Sheet into the Strath Glass Fault. Like many such faults in the district it is marked by topographic depressions and fault-line scarps and by a zone of breccia with carbonate, in places, over 10 m across. As this fault dips steeply north (about 70° at one locality) and transects strata dipping steeply but not vertically it may be a normal or oblique-slip rather than a transcurrent fault.

Another group of faults, some of which show evidence of transcurrent movement (for instance C on Figure 47) trend in a north-westerly direction. These, together with the parallel joints, have in some cases been invaded by Devonian lamprophyre intrusions. On the Kintail (72W) Sheet it can be shown that some of the NW-trending structures were initiated before much of the movement on the Strathconon fault complex, as felsite dykes intruded on them have been sinistrally displaced. West and south of Loch Cluanie there are two dextral faults, the more southerly of which can be traced north-westwards for about 8 km on the adjoining sheet where it has a displacement of about 0.6 km. At D on Figure 47 it displaces an ENE-trending sinistral fault and the orientation of associated joint drag folds suggests that at this locality it is an oblique slip fault with downthrow to the north.

East of Loch Cluanie there are several ENE- and NE-trending faults, two of which cut and therefore postdate the Ceannacroc breccias (Figure 43). Though two of these faults have large apparent lateral displacements, they cut strata dipping at less than 60° and could be normal or oblique slip faults.

MINERALISATION

Galena and barytes occur in specimen quantities in an ENE-trending crush zone 0.5 m wide in the Allt nam Peathrain [191 109] NE of the Cluanie dam (Figure 47). At a locality [109 029] on the Lochy Lochy (62E) Sheet just outside the Glen Affric district a carbonate-cemented breccia in an ENE-trending fault contains very small quantities of galena and sphalerite. The ENE-trending fault shown on the Loch Lochy Sheet a few metres north of the latter locality contains a brecciated camptonite dyke, suggesting that the carbonate-galena-sphalerite mineralisation may be associated with this intrusive igneous suite.

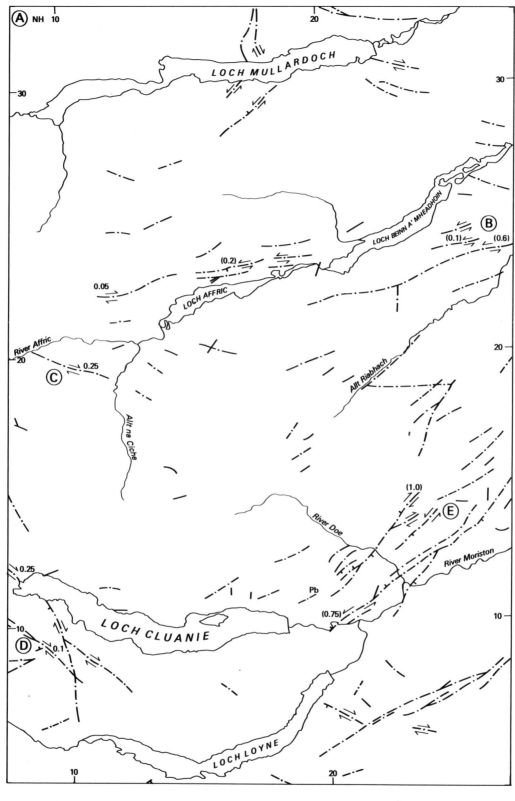

Figure 47   Faults and mineralisation.

NINE

Pleistocene and Recent

INTRODUCTION

The Pleistocene history of the area was one of repeated glaciation, separated by interstadial and interglacial periods when normal erosion and subaerial processes were active. This has resulted in the present landscape dominated by U-shaped valleys and corries. During the maximum of the last (Devensian) glaciation, about 18 000 BP (radiocarbon-years ago), the area was under an ice-sheet which covered most of the Scottish Highlands. The ice-shed lay near the western margin of the map. Subsequently there is evidence for a period of valley glaciation, but it is not known whether this was merely consequent on the shrinkage of the ice-sheet or was a discrete event. Much of the area was, however, influenced by an advance or readvance of ice between 11 000 and 10 000 BP. This has been termed the Loch Lomond Readvance (the Loch Lomond Advance by some authors) and most of the deposits and many small-scale features of glacial erosion preserved today were formed during this period. In postglacial times (between 10 000 BP and the present), the deposits left by the glaciers were partly eroded and redistributed by hillwash, landslips and rivers.

SUPERFICIAL DEPOSITS

The glacial deposits of the area include till and loose morainic debris, which in places are interbedded with or overlain by water-sorted gravel, sand, silt and clay deposited by meltwater. These are generalised, together with the usually small stretches of alluvium, on Figure 48.

Till

Till is mainly confined to the low ground, where it may reach a thickness of several metres; good natural exposures of it are uncommon. Outwith the limit of the Loch Lomond Readvance (Figure 50) on the south slope of Glen Moriston and north of Loch Beinn a'Mheadhoin, much of the glacial drift is sandy till which ranges from loosely compacted to stiff and occurs as dissected sheets. Similar till is well exposed along forestry roads in Glen Affric within the readvance limit. At some localities, for instance at [183 147] on the north bank of the River Doe and at [244 253] on the south-east shore of Loch Beinn a'Mheadhoin, the basal till is packed with angular blocks derived from the underlying rock. On the north side of Loch Beinn a'Mheadhoin at [220 245], a roadside section exhibits a heterogeneous, crudely bedded assemblage of sandy till, gravel, sand and stony silt overlain by packed angular boulders clearly derived and transported from the same rock outcrop (Plate 7). An early stage in the derivation of such debris is illustrated by a loch-side exposure at [213 316], west of the north end of the Loch Mullardoch dam, where a striated whaleback surface has been broken and partly displaced, possibly as the result of stress-relief during unloading (see below, p.71).

Morainic drift

The till sheets pass laterally into looser morainic debris, which is locally of up to 10 m thick, as in upper Gleann Fada at the head of the River Doe drainage. Morainic debris forms smooth dissected sheets as well as mounds. Quarries and forestry road cuttings in such mounds on the south side of Loch Affric expose very heterogeneous material, chiefly water-washed but very poorly sorted silty sandy gravel and silty sand with numerous wisps of fine-grained sand. There are also thin beds of till-like material (diamicton). Bedding is mainly horizontal or parallel to the slope of the mound, but is locally steeply dipping and faulted. An exposure on the north side of the Glen Affric valley [214 240] shows vertical beds of laminated and gravelly sand interbedded with sandy diamicton (Plate 8). The attitude of the beds here is thought to be due to deposition on top of glacier ice, with subsequent collapse of the sediments when the ice melted. On the north side of Loch Cluanie, the following section, which occurs in gently rolling moundy drift by the lochside [121 109], may have a similar origin:

	Thickness m
Sand, fine- to coarse-grained, with silt forming lenses and discontinuous beds. Interbeds of sandy, matrix-supported gravel. Numerous subangular to subrounded clasts of pebble to boulder size	2.0
Gravel, fine- to coarse-grained, chiefly matrix-supported, with scattered cobbles and boulders, the latter (up to 0.5 m) partly bedded in the underlying bed	0.3
Gravel, matrix-supported, with a continuous thin bed of laminated silt and sand. Matrix of the gravel is a poorly sorted sand	1.0

Dips in these beds are chiefly subhorizontal, but locally vertical.

An adjacent section is shown in Figure 49. Here the collapse features seem to be confined to the top of the section. The lower material may be immature basal till with streams of dark and pale fragments derived from nearby bedrock.

It has been suggested that hummocky morainic drift like that in the Glen Affric district was initially derived from the exposed valley sides by severe periglacial weathering and then transported and deposited as supraglacial debris by active valley glaciers during the Loch Lomond stage (Eyles, 1983a). However, the large volume of drift involved, its local origin, and the large size of the postulated glaciers compared to the relatively small area of exposed ground suggests that

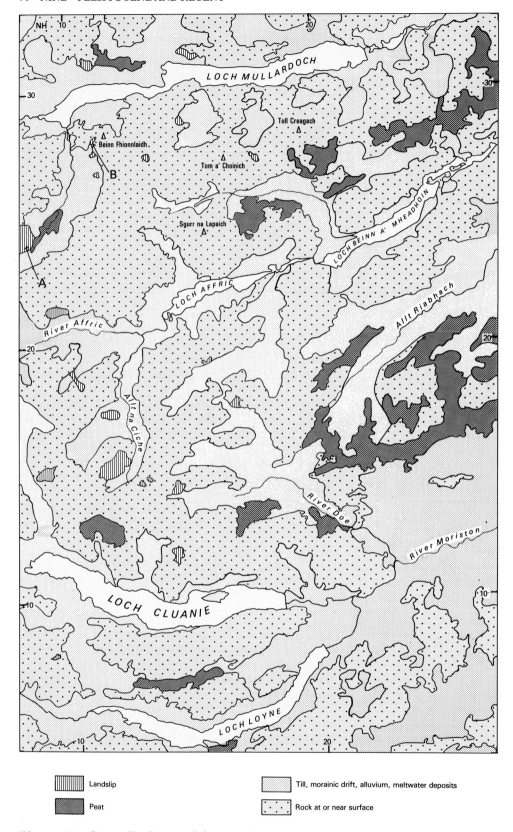

Figure 48 Generalised map of the superficial deposits.

Plate 7 Boulders lodged by ice moving eastwards (to right of picture). Till and sand below. Glaciated pavement near right side of exposure. Road section on north side of Loch Beinn a'Mheadhoin [220 245]. D2267.

subglacially derived debris may form an important fraction of morainic drift in the Glen Affric district (cf. Bailey and Maufe, 1960, p.276). The basal debris may originally have been formed in part from fracturing of the rock surface consequent on stress relief during unloading as the main Devensian ice sheet retreated (see Eyles and Paul, 1983, p.124). It was then incorporated into the basal load of the readvance glacier. A contribution to the basal debris may also have been made by a pre-existing regolith or drift sheet.

Terminal moraines attributed to the Loch Lomond stade in Glen Moriston and Glen Affric are for the most part ridges and mounds 1 to 5 m high; these are formed of poorly sorted gravel and boulders. Cross-valley moraines in the valley of the River Doe (Figure 51) seem to consist of both sand and till.

Features in drift and rock aligned in the direction of ice movement occur in the south part of the area (Peacock, 1967b), particularly in the valley of the River Loyne, where there are till ridges up to 1.5 km long. Such ridges, though visible on air photographs, are often difficult to see on the ground.

Meltwater deposits

Bedded deposits laid down by glacial meltwaters either as proglacial sheets or in contact with glacier ice occur widely, but significant deposits are confined to Glen Moriston, the valley of the River Doe and its tributaries, and the valley of the Allt Riabhach. In Glen Moriston, meltwater deposits occur chiefly in the valley bottom as far east as Dundreggan reservoir (Sheet 73W). They comprise gently sloping outwash and moundy, ice-contact sand and gravel deposited during the retreat of the Loch Lomond Readvance glacier from its maximum extent at the position of the present reservoir. A fine esker occurs by the roadside east of Dundreggan Lodge [Sheet 73W 335 147]. In the valley of the River Doe and its tributaries there are deltaic and lacustrine silts, clays, fine-grained sands and gravels up to 30 m thick, deposits of a former glacial lake (Figure 51). Further north, in the valley of the Allt Riabhach, moundy sands and gravels and outwash terraces extend north-eastwards from Garve Bridge [267 223] for some 6 km to Tomich (Figure 50). Elsewhere in this more northern part of the Sheet, small volumes of ter-

Plate 8 Steeply dipping interbedded sandy diamicton (till) with sandy lenticles, gravel and laminated sand. Section in morainic mound on forestry road near junction of Gleann nam Fiadh and Glen Affric [214 240]. D2264.

race and moundy sand and gravel occur in Gleann nam Fiadh and the valley of the Allt na h-Imrich, north of Loch Beinn a'Mheadhoin. There is a system of small sandy eskers in the valley of the Allt Garbh [172 195].

Periglacial deposits

Such deposits, which are partly late-glacial and partly of more recent age, occur particularly on the higher hills in the west half of the Glen Affric district, where solifluction terraces, blockfields (felsenmeer) and obliquely sloping terracettes formed by wind action are common. The area outside the presumed limit of the Loch Lomond Readvance (see below) appears to have been subjected to more mass movement than that within it and there the slopes of morainic landforms tend in some cases to be more subdued. Stone lobes, possibly formed during the cold period of the Loch Lomond stade, occur sparsely. They are developed, for instance, on the north slope of Toll Creagach [194 283]. Small moraine-like knolls at the northern base of the ridge [172

268] extending south-east from Tom a'Choinich may be fossil protalus ramparts formed at the base of former snow slopes.

Landslips

Most landslips (Figure 48) have taken place in the area of higher, steeper ground in the western half of the map area, especially on southern and eastern hill slopes. In some cases incipient slips are revealed as open fractures in otherwise undisturbed ground, for instance at [246 274] on the south slope of Meall Mor. The incipient slip on the north side of Sgurr na Lapaich [154 244] is backed by a low scarp on the summit ridge which can be traced for over 0.5 km. It is probable that many slips originated in late-glacial times during ice retreat as a result of the release of stress combined with a high water table (Eyles, 1983b, p.104) in areas where there is a favourable attitude of joints and foliation. A few slips may still be active, for instance parts of the very large landslip on the west slope of Glean a' Choilich (A on Figure 48).

N   S

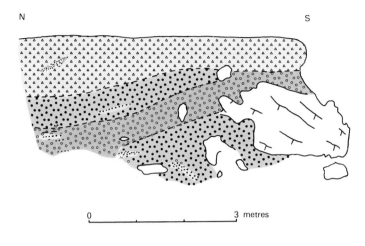

0   3 metres

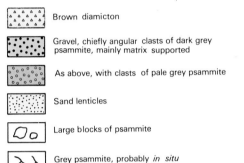

Brown diamicton

Gravel, chiefly angular clasts of dark grey
psammite, mainly matrix supported

As above, with clasts of pale grey psammite

Sand lenticles

Large blocks of psammite

Grey psammite, probably *in situ*

Figure 49 Vertical section seen in a morainic mound on the north side of Loch Cluanie.

Scallops and corries on the west side of this valley could be the sites of former landslips removed during glaciation (cf. Clough in Gunn et al., 1897, p.276). A nearby slip on the west side of Beinn Fhionnlaidh [113 283] is accompanied by a debris flow (B on Figure 48).

Peat and alluvium

Blanket peat is common in much of the area. It is particularly widespread on the ground of lower relief in the east where the deposit covers much of the gently sloping watershed between the Glen Moriston and Glen Affric drainages (Figure 48). The peat is locally over 3 m thick. Small stretches of alluvium (gravel, sand and silt) margin some of the streams, passing outwards at some localities into peat bogs. Larger areas of alluvium in the major valleys commonly consist of redistributed glacial gravels.

LATE-GLACIAL HISTORY

Events predating the Loch Lomond Stade

Evidence relating to this period is found outwith the readvance limits (Figure 50) in Glen Affric and Glen Moriston. In Fasnakyle Forest, between Glen Cannich and Loch Beinn

a'Mheadhoin, striated surfaces above 800 m above OD record the eastward flow of ice during the main Devensian glaciation. This ice sheet has left a sheet of till on the south slopes of Glen Moriston. The area with little till cover between Glen Affric and Glen Moriston passes eastward into well-developed crag and tail features south-west of Glen Urquhart on the adjacent Invermoriston (73W) Sheet.

In the valley of the River Doe there is evidence for a period of valley glaciation which postdated the main Devensian ice but which predated the Loch Lomond Readvance. From the evidence of striations and the distribution of erratics of the Cluanie Granodiorite, ice flowed north-eastwards out of Glen Moriston across the Carn nam Feuaich ridge (732 m above OD) and the Doe valley (Figures 50 and 51). It thus followed much the same course as the (later) Loch Lomond Readvance glacier. Boulders of granodiorite extend up to at least 560 m above OD on the north slope of the River Doe valley, above the presumed limit of the later readvance glacier and its accompanying glacial lakes. Retreat features associated with this earlier valley glaciation include a well marked upper limit of boulders at about 595 m above OD on the south side of An Reithe [160 148] in the upper Doe drainage area (Figure 51). This was taken as the upper limit for the Loch Lomond Readvance by Peacock (1975), but subsequent work suggests that this feature is probably outside the limit (Sissons, 1977). Other features include a belt of end or lateral moraine at about 430 m above OD on the east side of lower Gleann Fada and an area of kettled boulder gravel on the north side of Carn nam Feuaich (Figure 51). On the north side of Glen Moriston there are meltwater channels above the limit of the Loch Lomond Readvance ice (Figure 51).

Loch Lomond Readvance

The glaciers of this episode (also termed the Loch Lomond Advance by some workers) left well-marked terminal features in Glen Moriston and Glen Affric (Figure 50). These have been correlated on geomorphological grounds with glaciers of the Loch Lomond Stade elsewhere (Peacock, 1975; Sissons, 1977). The limit of the glacier in Glen Cannich is much less clear. In places, such as the south side of Glen Moriston (Sissons, 1977) and in Gleann nam Fiadh, multiple terminal moraines suggest that the glaciers remained near the readvance limit for more than a few years. Within the present area no direct stratigraphical or lithological evidence for a readvance, such as the overriding and incorporation of earlier, dateable deposits, is known. The readvance limit is defined solely on geomorphological criteria such as moraine ridges and meltwater channels and by the presence of periglacial features (boulder lobes, large solifluction terraces and extensive frost-shattered rock) which tend to be well developed outside the limit but less so within it. However, the extent of frost shattering and other periglacial phenomena is strongly dependent on lithology, being much greater in quartzite and psammite than in mica-schist. It is, therefore, not always feasible to use these as guides to the presumed limit of the readvance in an area of complex geology such as the Glen Affric district.

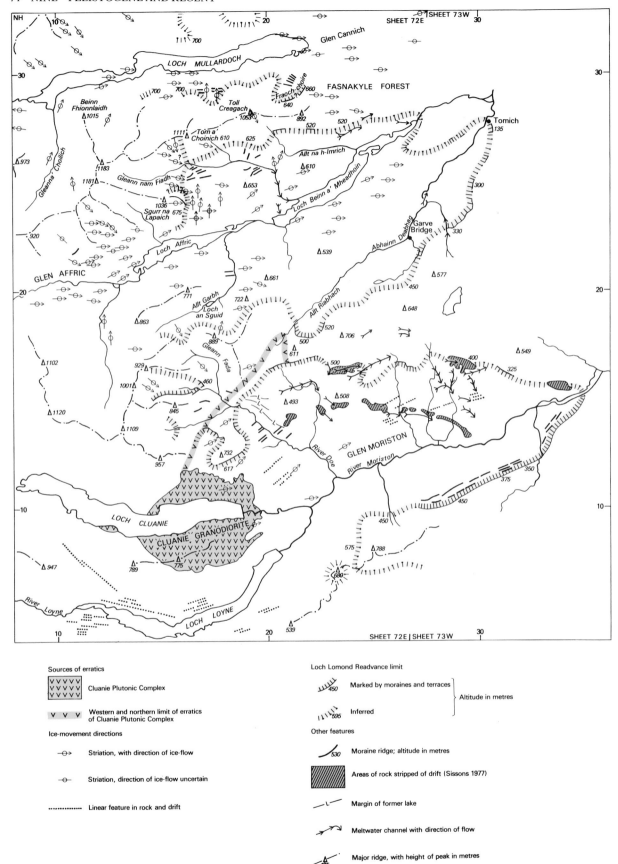

Figure 50 Glacial features and Loch Lomond Readvance limit in the Glen Affric district.

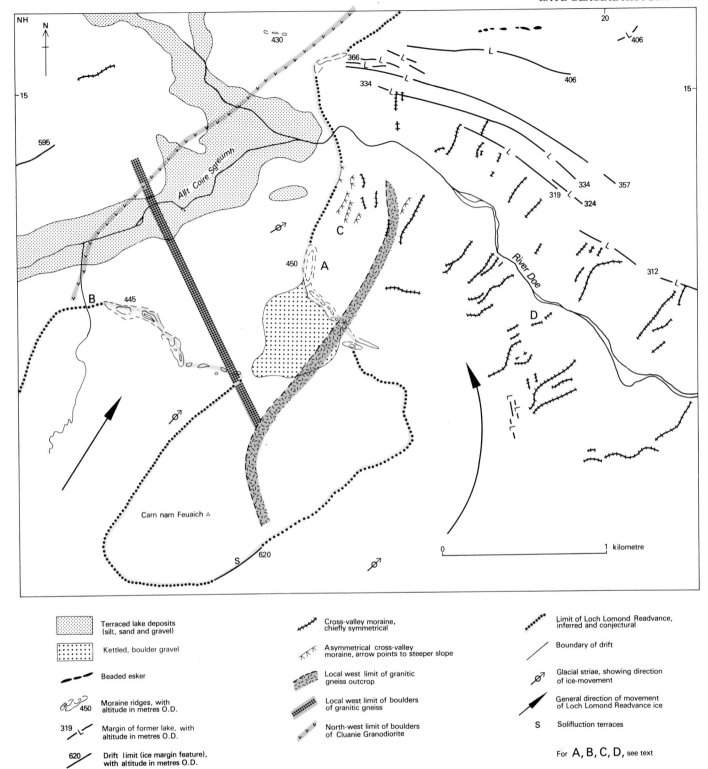

Figure 51 Glacial and pro-glacial features of the River Doe area.

Glen Cannich

A morainic terrace associated with features resembling cross-valley moraines can be seen in Froach-choire, south of the Mullardoch dam (Figure 50). This could be either a terminal feature of the Loch Lomond Readvance or an earlier feature of valley glaciation. No supporting evidence for the former

view, such as a contrast between periglacial features within and without the limit, has been found. However, the probability that a glacier of the Loch Lomond Stade occupied at least the upper half of this glen is suggested by the presence of fresh-looking moraine mounds, the absence of evidence of corrie glaciation, and the position of such a glacier relative to the large valley glacier of this age in Glen Affric. A tentative

Plate 9 Terminal/lateral moraine complex attributed to Loch Lomond Readvance glacier in Glen Affric. Smooth slope of till and peat to left. North side of Gleann nam Fiadh [197 265]. D2256.

limit for the readvance ice based on relative frost shattering is shown on Figure 50.

Glen Affric

In this valley and in the valley of the Abhainn Deabhag to the south, the limit of the readvance is marked by moraines, channels and terraces, which are particularly well seen on the north side of Glen Affric (Figure 50). South-west of Tomich a terminal moraine and associated meltwater channels and deposits can be traced in the vicinity of the forestry fence for about 2.5 km. On the hillside above the moraine there is bare rock and thin drift, but below it there are numerous meltwater channels and small terraces that slope obliquely across the hillside. Tomich itself is at the upstream limit of gravel terraces deposited beyond the margin of the Loch Lomond Readvance glacier, and the ice limit here is further indicated by moraines north and south of the village. Near the head of the Abhainn Deabhag, cols into the Glen Moriston drainage area seem to have been the sites of short-lived glacial lakes (Sissons, 1977) for instance at [270 170].

North of Loch Beinn a'Mheadhoin, lateral moraines associated with terraces and meltwater channels can be traced for nearly 10 km. These form the northern limit of the Loch Lomond Readvance glacier in Glen Affric. Above this is a smooth slope of till and peat (Plate 9). Westwards, the limit can be traced across Gleann nam Fiadh, the lower part of which was blocked by ice. The extent of ice in the upper part of this glen is uncertain, but a small glacier seems to have descended Coir Mhic Fhearchair [165 267] south-west of Tom a'Choinich, as shown by fresh-looking striae on exposed surfaces of psammite. The series of terminal moraines of the main Glen Affric glacier north-east of Sgurr na Lapaich are composed largely of psammite blocks derived from the south. Cross valley moraines located just above the major bend in Gleann nam Fiadh could have formed at the ice margin in an upstream lake dammed by the glacier (see discussion of the Glen Moriston area) but there is no evidence for an extensive body of water. Subglacial drainage evidently took place eastwards into the valley of the Allt na h-Imrich and southwards along lower Gleann nam Fiadh to form the gorges in both these glens.

Glen Moriston

Details pertaining to the advance and retreat of the Loch Lomond Readvance glacier here and in Glen Doe (Figures 50 and 51) are given by Sissons (1977). In this area, ice emanating from the Cluanie basin was contiguous on its south side with glaciers in the Loyne valley and Glen Garry (Sheet 62E). On the south side of Glen Moriston, the presumed readvance limit is marked by the upper limit of moundy drift and by a prominent outer moraine ridge in the area east of the Glen Affric Sheet (Figure 50). On the north side of the valley the limit is delineated by moraines, terraces, meltwater channels and by an upper limit of rock stripped bare of drift by marginal and submarginal floods of meltwater derived from the catastrophic draining of glacial lakes farther west (Sissons, 1977). The long, broad meltwater channels below the readvance limit here are also thought by Sissons to have been associated with such catastrophic drainage. The lower belt of bare rock may define a later position of the ice front at a time when glacial lakes were still present farther west in the Doe valley.

In the valleys which form the drainage system of the River Doe there were several glacier tongues (Figure 50), the positions of which have been determined from end moraines, upper limits of moundy moraines and on high ground by contrasts in the degree of frost shattering (slight within the limit, marked outside it). The glaciers terminated in a lake, dammed up by a tongue of ice extending into the Doe valley from Glen Moriston. The almost straight ice fronts in this type of situation are inferred from moraines and other features such as a sharp up-valley limit of a drift sheet near the head of the River Doe (Figure 51). The erratics of Cluanie Granodiorite in the bottom of the River Doe valley upstream of the readvance limit might be related locally to transport by icebergs in the lake, the level of which at one time reached to a little above 500 m above OD (Sissons, 1977). The boulder moraine marking the readvance limit on the south side of the Doe valley (A in Figure 51), some 60 m across and 8 m high, owes its dimensions to the fact that it was formed largely from pre-existing kettled drift (see above).

The upper Doe valley is floored by lacustrine and deltaic deposits, chiefly fine-grained sand with subsidiary coarse sand and gravel, which were dissected and terraced during the fall of water level. The instability of slopes in such fine-grained deposits is demonstrated by numerous slumps and mudflows. Contemporaneity of the lake deposits with the formation of the terminal moraines is shown by the interbedding of laminated silt, sand and clay with morainic debris at the localities where glacier tongues entered the lake, for instance at B in Figure 51.

Lake shorelines that were cut in poorly sorted sandy bouldery drift are prominent features on the north side of the Doe valley, where 10 levels of 'parallel roads' have been mapped, whereas on the south side of the valley only 3 have been seen. Shoreline levels at about 406, 366, 357, 334, 324, 319 and 312 m above OD have been recorded by instrumental means and by interpolation on the north side of the valley. On the south side of the valley, the three features seem to be at 340, 334 and 324 m, but have not been levelled. Of the shorelines on the north side those at 406 m above OD (tread width of between 3 and 13 m) and 357 m (tread width up to 5 m) are the most clear cut. The 406 m shoreline corresponds to a broad meltwater channel starting at this level and to the lower area of water-washed rock stripped of drift (Figures 50 and 51). Thus the Glen Moriston glacier could have retreated nearly 6 km before the lake at this level was drained. Details of the drainage paths of some of the lower lakes are given by Sissons (1977).

The retreating of the ice front eastwards left a series of cross-valley moraines on the floor of Glen Doe. Those near the readvance limit (C on Figure 51) are only about a metre or so high and are asymmetric (steeper distally). They may be push-features formed annually by the forward movement of the sole of the glacier in winter. The moraines farther down-valley range up to 15 m in height and are more or less symmetrical, as noted by Sissons (1977). The parts of the latter ridges adjacent to the river are apparently formed chiefly of unbedded sand with a few boulders, but they may pass up-slope into features composed of till. At one locality (D on Figure 51) sections show that the sand forming a mound passes downwards into apparently undisturbed, interbedded gravel and laminated silt and sand some 3 to 6 m thick which rest in turn on till. There is as yet no satisfactory hypothesis for the generation of these generally higher ridges (Sissons, 1977), though they may also have been formed at or near the ice front, perhaps in part from debris falling from an undercut ice cliff above and in part from lake sediment reworked by the ice.

REFERENCES

Most of the references listed below are held in the Library of the British Geological Survey at Keyworth, Nottingham. Copies of the references can be purchased subject to the current copyright legislation.

BAILEY, E B and MAUFE, H B. 1960. The geology of Ben Nevis and Glen Coe. *Memoir of the Geological Survey of Great Britain* (2nd edition).

BOWES, D R, MACDONALD, G D, MATHESON, G D, and WRIGHT, A E. 1963. An explosion-breccia appinite complex at Gleann Charnon, Argyll. *Transactions of the Geological Society of Glasgow*, Vol. 25, 19–30.

BOWES, D R, and WRIGHT, A E. 1967. The explosion-breccia pipes near Kentallen, Scotland, and their geological setting. *Transactions of the Royal Society of Edinburgh*, Vol. 67, 109–143.

BROOK, M, POWELL, D, and BREWER, M S. 1976. Grenville age for rocks in the Moine of northwestern Scotland. *Nature London*, Vol. 360, 515–517.

BROOK, M, POWELL, D, and BREWER, M S. 1977. Grenville events in Moine rocks of the Northern Highlands, Scotland. *Quarterly Journal of the Geological Society of London*, Vol. 133 489–496.

CLIFFORD, T N. 1958. The stratigraphy and structure of part of the Kintail district of Southern Ross-shire: its relation to the Northern Highlands. *Quarterly Journal of the Geological Society of London*, Vol. 113, 57–85.

DALZIEL, I W D. 1966. A structural study of the granitic gneiss of Western Ardgour, Argyll and Inverness-shire. *Scottish Journal of Geology*, Vol. 2, 125–152.

DEARNLEY, R. 1967. Metamorphism of minor intrusions associated with the Newer Granites of the Western Highlands of Scotland. *Scottish Journal of Geology*, Vol. 3, 449–457.

EYLES, N. 1983a. Modern Icelandic glaciers as depositional models for 'hummocky moraine' in the Scottish Highlands. 47–59 in *Tills and related deposits*. EVENSON, E B, SCHLUCHTER, C, and RABASSA, J (editors). (Rotterdam: Balkema.)

— 1983b. The glaciated valley land system. 91–110 in *Glacial geology*. EYLES, N (editor). (Oxford: Pergamon Press.)

— and PAUL, M A. 1983. Landforms and sediments resulting from former periglacial climates. 111–139 in *Glacial geology*. EYLES, N (editor). (Oxford: Pergamon Press.)

FETTES, D J, and MACDONALD, R. 1978. Glen Garry Vein Complex. *Scottish Journal of Geology*, Vol. 14, 335–358.

— LONG, C B, BEVINS, R E, MAX, M D, OLIVER, G J H, PRIMMER, T J, and THOMAS, L J. 1984. Grade and time of metamorphism of the Caledonide Orogen of Britain and Ireland. *Memoir of the Geological Society of London*, No. 9.

GEOLOGICAL SURVEY OF GREAT BRITAIN 1965. *Summary of progress for 1964.* (London: Her Majesty's Stationery Office.)

GODARD, A. 1965. *Recherches de géomorphologie en Ecosse du Nord-Ouest.* (Paris: University of Strasbourg.)

GUNN, W, CLOUGH, C T, and HILL, J B. 1897. The geology of Cowal. *Memoir of the Geological Survey of Great Britain.* 333 pp.

HARRIS, A L. 1983. The growth and structure of Scotland. 1–22 in *Geology of Scotland.* CRAIG, G Y (editor). (Edinburgh: Scottish Academic Press.)

HOLDSWORTH, R E, and ROBERTS, A M. 1984. A study of early curvilinear fold structures and strain in the Moine of the Glen Garry region, Inverness-shire. *Quarterly Journal of the Geological Society of London*, Vol 141, 327–388.

HORNE, J, and HINXMAN, L W. 1914. The geology of the country round Beauly and Inverness. *Memoir of the Geological Survey of Great Britain.*

JOHNSTONE, G S. 1989. *British regional geology: the Northern Highlands* (4th edition). (Edinburgh: HMSO for British Geological Survey.)

— SMITH, D I, and HARRIS, A L. 1969. The Moinian Assemblage of Scotland. 159–180 in North Atlantic geology and continental drift. KAY, M (editor). *American Association of Petroleum Geologists*, Vol. 12.

LEEDAL, G P. 1953. The Cluanie igneous intrusion, Inverness-shire and Ross-shire. *Quarterly Journal of the Geological Society of London*, Vol. 108, 35–63.

MACGREGOR, A G. 1948. Resemblances between Moine and 'Sub-Moine' metamorphic sediments in the Western Highlands of Scotland. *Geological Magazine*, Vol. 83, 265–275.

MOORHOUSE, S J, and MOORHOUSE, V E. 1979. The Moine amphibolite suites of central and eastern Scotland. *Mineralogical Magazine*, Vol. 43, 211–225.

NORTH OF SCOTLAND HYDRO-ELECTRIC BOARD. 1962–1966. *Report and accounts for the years 1961–1964 and 1st April to 31 March 1966.* (Edinburgh: HMSO.)

PEACH, B N, GUNN, W, CLOUGH, C T, HINXMAN, L W, CRAMPTON, C B, and ANDERSON, E M. 1912. The geology of Ben Wyvis, Carn Chuinneag, Inchbae and the surrounding country. *Memoir of the Geological Survey of Great Britain.*

PEACOCK, J D 1967. West Highland morainic features aligned in the direction of ice flow. *Scottish Journal of Geology*, Vol. 3, 372–373.

— 1972. Sodic rocks of metasomatic origin in the Moine Nappe (letter). *Scottish Journal of Geology*, Vol. 8, No. 3, 291–292.

— 1975. Palaeoclimatic significance of ice-movement directions of Loch Lomond readvance glaciers in the Glen Moriston and Glen Affric areas, northern Scotland. *Bulletin of the Geological Survey of Great Britain*, No. 49, 39–42.

— 1977. Metagabbros in granitic gneiss, Inverness-shire, and their significance in the structural history of the Moines. *Report of the Institute of Geological Sciences*, No. 77/20.

PEARCE, J A, and NORRY, M J. 1979. Petrogenetic implications of Ti, Zr, Y and Nb variations in volcanic rocks. *Contributions to Mineralogy and Petrology*, Vol. 69, 33–47.

PIDGEON, R T, and AFTALLION, M. 1978. Cogenetic and inherited zircon U-Pb systems in granites: Palaeozoic granites of Scotland and England. 183–220 in *Crustal evolution in northwestern Britain and adjacent regions*. BOWES, D R, and LEAKE, B E (editors). (Liverpool: Seel House Press.)

POWELL, D. 1983. Time of deformation in the British Caledonides. 293–300 in Regional trends in the geology of the Appalachian–Caledonian–Hercynian–Mauritanide Orogen. SCHENK, P E (editor). *NATO ASI Series C Mathematical and Physical Sciences*, Vol. 116 (Dordrecht: D. Reidel Publishing Co.)

RAMSAY, J G. 1955. A camptonite dyke suite at Monar, Ross-shire and Inverness-shire. *Geological Magazine*, Vol. 92, 297–308.

— 1958. Superimposed folding at Loch Monar, Inverness-shire and Ross-shire. *Quarterly Journal of the Geological Society of London*, Vol. 113, 271–307.

RATHBONE, P A, and HARRIS, A L. 1979. Basement-cover relationships at Lewisian inliers in the Moine rocks. *In* The Caledonides of the British Isles—reviewed. HARRIS, A L, HOLLAND, C H, and LEAKE, B E (editors). *Special Publication of the Geological Society of London*, No. 8.

READ, H H, PHEMISTER, J, and ROSS, G. 1926. The geology of Strath Oykell and lower Loch Shin. *Memoir of the Geological Survey of Scotland*.

RICHEY, J E, and KENNEDY, W Q. 1939. The Moine and Sub-Moine Series of Morar, Inverness-shire. *Bulletin of the Geological Survey of Great Britain*, Vol. 2, 26–45.

ROBERTS, A M, and HARRIS, A L. 1983. The Loch Quoich Line—a limit of crustal reworking in the Northern Highlands of Scotland. *Journal of the Geological Society of London*, Vol. 140, 883–892.

— SMITH, D I, and HARRIS, A.L. 1984. The structural setting and tectonic significance of the Glen Dessary Syenite, Inverness-shire. *Quarterly Journal of the Geological Society of London*, Vol. 141, 1033–1042.

ROCK, N M S. 1982. The Permo-Carboniferous camptonite-monchiquite dyke-suite of the Scottish Highlands and Islands: distribution, field and petrological aspects. *Report of the Institute of Geological Sciences*, No. 82/14.

— 1983. The geology of the area around Cannich, Inverness-shire. *Institute of Geological Sciences, Petrographical Report*, No. 5022.

— 1984. New types of hornblendic rocks and prehnite-veining in the Moine west of the Great Glen, Inverness-shire. *Report of the Institute of Geological Sciences*, No. 83/8.

— MACDONALD, R, WALKER, B H, MAY, F, PEACOCK, J D, and SCOTT, P. 1985. Intrusive metabasite belts within the Moine assemblage west of Loch Ness, Scotland: evidence for metabasite modification by country rock interactions. *Quarterly Journal of Geological Society of London*, Vol. 142, 643–662.

SIMONY, P. 1963. Structural and metamorphic geology of the Saddle area, Wester Ross and Inverness. Unpublished PhD thesis, University of London.

SISSONS, J.B. 1977. Former ice-dammed lakes in Glen Moriston, Inverness-shire, and their significance in upland Britain. *Transactions of the Institute of British Geographers*, New Series, Vol. 2, 224–242.

SMITH, D I. 1979. Caledonian minor intrusions in the N. Highlands of Scotland. 683–697 *in* The Caledonides of the British Isles—reviewed. HARRIS, A L, HOLLAND, C H, and LEAKE, B E (editors). *Special Publication of the Geological Society of London*, No. 8.

SPRING, J, and RAMSAY, J G. 1962. Moine stratigraphy in the western Highlands. *Proceedings of the Geologists Association*, Vol. 73, 295–322.

TALBOT, C J. 1983. Microdiorite sheet intrusions as incompetent time- and strain markers in the Moine assemblage NW of the Great Glen fault, Scotland. *Transactions of the Royal Society of Edinburgh*, Vol. 74, 137–152.

TANNER, P W G. 1976. Progressive regional metamorphism of thin calcareous bands from the Moinian rocks of NW Scotland. *Journal of Petrology*. Vol. 17, 100–134.

— JOHNSTONE, G S, SMITH, D I, and HARRIS, A L. 1970. Moinian stratigraphy and the problem of the Central Ross-shire Inliers. *Bulletin of the Geological Society of America*, Vol. 81, 299–306.

— and TOBISCH, O T. 1972. Sodic and ultra-sodic rocks of metasomatic origin from part of the Moine Nappe. *Scottish Journal of Geology*, Vol. 8, 151–178.

TOBISCH, O T. 1963. The structure and metamorphism of the Moinian rocks in Glen Cannich—Fasnakyle Forest area. Unpublished PhD thesis, University of London.

— 1965. Observations on primary deformed sedimentary structures in some metamorphic rocks from Scotland. *Journal of Sedimentary Petrology*. Vol. 35, 415–419.

— 1966. Large-scale basin-and-dome pattern resulting from the interference of major folds. *Bulletin of the Geological Society of America*, Vol. 77, 393–408.

— 1967. The influence of early structures on the orientation of late-phase folds in an area of repeated deformation. *Journal of Geology*, Vol. 75, 554–564.

— FLEUTY, M J, MERH S S, MUKHOPADHYAY, D, and RAMSAY, J G. 1970. Deformational and metamorphic history of Moinian and Lewisian rocks between Strathconon and Glen Affric. *Scottish Journal of Geology*, Vol. 6, 243–265.

WINCHESTER, J A. 1972. The petrology of Moinian calc-silicate gneisses from Fannich Forest, and their significance as indicators of metamorphic grade. *Journal of Petrology*, Vol. 13, 405–424.

— 1974. The zonal pattern of metamorphism in the Scottish Caledonides. *Quarterly Journal of the Geological Society of London*, Vol. 130, 509–524.

— 1976. Different Moinian amphibolite suites in northern Ross-shire. *Scottish Journal of Geology*, Vol. 12, 187–204.

— and FLOYD, P A. 1977. Geochemical discrimination of different magma series and their differentiation products using immobile elements. *Chemical Geology*, Vol. 20, 325–343.

— LAMBERT, R St J, and HOLLAND, J G. 1981. Geochemistry of the western part of the Moinian assemblage. *Scottish Journal of Geology*, Vol. 17, 281–294.

INDEX

BRITISH GEOLOGICAL SURVEY

Keyworth, Nottingham NG12 5GG
(0602) 363100

Murchison House, West Mains Road, Edinburgh
EH9 3LA   031-667 1000

London Information Office, Natural History Museum
Earth Galleries, Exhibition Road, London SW7 2DE
071-589 4090

The full range of Survey publications is available
through the Sales Desks at Keyworth and at Murchison
House, Edinburgh, and in the BGS London
Information Office in the Natural History Museum
Earth Galleries. The adjacent bookshop stocks the
more popular books for sale over the counter. Most
BGS books and reports are listed in HMSO's Sectional
List 45, and can be bought from HMSO and through
HMSO agents and retailers. Maps are listed in the BGS
Map Catalogue, and can be bought from Ordnance
Survey agents as well as from BGS.

*The British Geological Survey carries out the geological survey
of Great Britain and Northern Ireland (the latter as an
agency service for the government of Northern Ireland), and
of the surrounding continental shelf, as well as its basic
research projects. It also undertakes programmes of British
technical aid in geology in developing countries as arranged
by the Overseas Development Administration.*

*The British Geological Survey is a component body of the
Natural Environment Research Council.*

HMSO publications are available from:

HMSO Publications Centre
(Mail, fax and telephone orders only)
PO Box 276, London SW8 5DT
Telephone orders 071-873 9090
General enquiries 071-873 0011
Queueing system in operation for both numbers
Fax orders 071-873 8200

HMSO Bookshops
49 High Holborn, London WC1V 6HB
(counter service only)
071-873 0011   Fax 071-873 8200
258 Broad Street, Birmingham B1 2HE
021-643 3740   Fax 021-643 6510
Southey House, 33 Wine Street, Bristol BS1 2BQ
0272-264306   Fax 0272-294515
9 Princess Street, Manchester M60 8AS
061-834 7201   Fax 061-833 0634
16 Arthur Street, Belfast BT1 4GD
0232-238451   Fax 0232-235401
71 Lothian Road, Edinburgh EH3 9AZ
031-228 4181   Fax 031-229 2734

HMSO's Accredited Agents
(see Yellow Pages)

And through good booksellers